Thomas Trotter

Medicina Nautica

An essay on the diseases of seamen: comprehending the history of health in His

Majesty's fleet, under the Command of Richard Earl Howe

Thomas Trotter

Medicina Nautica
An essay on the diseases of seamen: comprehending the history of health in His Majesty's fleet, under the Command of Richard Earl Howe

ISBN/EAN: 9783337412043

Printed in Europe, USA, Canada, Australia, Japan

Cover: Foto ©berggeist007 / pixelio.de

More available books at **www.hansebooks.com**

Medicina Nautica:

AN

ESSAY

ON

THE DISEASES OF SEAMEN:

COMPREHENDING

THE HISTORY OF HEALTH

IN

HIS MAJESTY's FLEET,

UNDER THE COMMAND OF

RICHARD EARL HOWE, ADMIRAL.

By THOMAS TROTTER, M. D.

MEMBER OF THE ROYAL MEDICAL SOCIETY;

AN HONORARY MEMBER OF THE ROYAL PHYSICAL SOCIETY, &C.

PHYSICIAN TO THE FLEET.

" *Grave Martis Opus.*"

ÆNEID.

LONDON:

PRINTED FOR T. CADELL, JUN. AND W. DAVIES, (SUCCESSORS

TO MR. CADELL) IN THE STRAND.

1797.

DEDICATION.

TO

RICHARD EARL HOWE,

Admiral:

THE FLAG OFFICERS,
CAPTAINS,
AND OTHER
OFFICERS OF THE FLEET.

MY LORDS AND GENTLEMEN,

DURING the unexampled Health of
the Fleet for fome time paft, and while
my attendance could be difpenfed with,
my leifure hours have been employed
in compiling this Work: I now humbly
folicit your patronage.

It

It comprehends the hiſtory of Health during a period of conſiderable exertion. To You, in a particular manner, theſe labours look for protection ; for You have witneſſed the whole.

It has been my wiſh, to make the duties of the Medical Profeſſion ſubſervient to the comfort and happineſs of Men, that have earned laurels for You, and given ſecurity to their Country. In doing this, I have been neither tenacious of form, or ſcrupulous about correcting old cuſtoms. As poſterity will only receive from us, what they may deem beneficial to themſelves ; ſo we have a right to reject or embrace the opinions, of thoſe who have gone before us, as may beſt ſuit our purpoſe. Medicine is connected with ſo many branches of ſcience, that as they improve, it is neceſſary to incorporate every

every difcovery into our fyftem of Health.
The LORDS COMMISSIONERS of ADMI-
RALTY, have bountifully attended to our
applications : Your teftimony of their
utility will, therefore, give frefh fupport to what we have further propofed.
When thefe improvements have received all the perfection of which they
are capable, we apprehend that they
will add refources to the naval power
of GREAT BRITAIN, that have never yet
been duly called to her affiftance. Thefe
are motives which urge me to lay this
Work before the Public, without delay : but it muft appear under all the
difadvantages infeparable from my prefent fituation. Whatever may be its
fate, here, or hereafter, it affords me
an opportunity, which I earneftly embrace, to return You my moft fincere
thanks, for the numerous inftances of
confidence,

confidence, friendſhip, and regard, which
have followed all my official and profeſſional engagements among You. The
impreſſion which they have made, ſhall
be treaſured in my heart; ſhall attend
me to retirement; and ſhall be cheriſh-
ed in remembrance, to the evening of
life.

Et moriens carum recordabor Argos.

I have the honour to be,

My LORDS and GENTLEMEN,

Your faithful humble Servant,

T. TROTTER.

Spithead, Dec. 12th
1796.

CONTENTS.

(viii)

ERRATUM.

Page 21, line 2, of the New Form, for *Welch linen*, read *Welch flannel.*

MEDICINA NAUTICA.

INTRODUCTION.

THE following pages contain the hiſtory of health, for three years, in a Fleet that has performed the moſt brilliant ſervices for Great Britain; which include the ſplendid victory of the 1ſt of June; the maſterly retreat of the "little ſquadron," under the Hon. Vice-admiral Cornwallis, from a fleet four times his force; and the handſome capture of three ſail of the line by Admiral Lord Bridport, in the very mouth of a French harbour.

The operations of a large Fleet in Channel ſervice offer a field for obſervation, of the firſt importance to the medical inquirer: that the preſent has afforded ſome novelty will ſcarcely be doubted. Ships newly commiſſioned, by receiving raw landmen, and drafts of ſeamen from crowded guard-ſhips, tenders, and hoſpitals, are the moſt liable to ſuffer from infection, and are generally employed, at firſt, on home-ſtations. Hence this

B   ſubject

subject has engaged much of our attention: but a
variety of other difeafes has prefented us with much
matter for animadverfion ; fuch as the fcurvy, catarrh, &c. I have endeavoured to imitate Dr.
Blane, in calling upon the furgeons for occafional
remarks; and much valuable information has been
received from that quarter, which I have thrown
into the form of notes.   It is to be regretted that
thefe communications were not more extenfive ;
but various caufes confpired to render an unin-
terrupted correfpondence impracticable : the chief
of thefe were, fhips being difpatched from the
fleet to different ports; the fluctuation of others,
and the defultory operations of the whole, after
the winter cruize, in February 1795.

My accounts of different difeafes, befides the
more immediate connection they have with naval
fervice, will be found ufeful to medical readers in
general.   They abound with facts that could be
met with no where elfe; and, being compiled from
a number of cafes, they bring into one view an
endlefs variety of fymptoms.   Of theories I have
faid little, or have only touched on fome, where
practical inductions rendered it unavoidable. We
have infufed into our fyftem of health fuch improvements, as our acquaintance with chemiftry,
in its cultivated ftate, warranted; this particularly applies to the means of fubduing contagion.
We have, therefore, entirely difregarded the
                                                            agents

agents employed by our predeceffors, for more than a century paft, as well as fome recent ones recommended by writers of the prefent day : and we contend that every fubftance whatever, that tends to diminifh the refpirable part of the atmofphere, fuch as the *gaffes*, in vogue, is hurtful in the extreme. We have not, however, thought it of fufficient moment to fearch the firft authorities for the employment of thefe fumes; they, doubtlefs, fprung from falfe hypothefis and inaccurate experiment, and have been purfued without fcrutiny. To a hyperoxygenated atmofphere, or one poffeffing its due proportion of oxygene, we look for fecurity againft infection; not as acting, by chemical combination, on contagious miafma; but as fupplying the human body with a quality that enables it to refift the offending power. But we do not reft this preference on an appeal to firft principles only; we have witneffed its fuccefs on a larger fcale of experience than has ufually fallen to the fhare of one obferver.

Our ftock of facts on the fubject of fcurvy is great beyond all precedent : its prevention and cure have, from recent experiment, been brought to a certainty, fo as to fuperfede the utility of future inveftigation.

In the treatment of typhus we alfo hope to have made improvements : from the advantage of

having

having attended an immenfe number of cafes in very diverfified fituations, we fuppofe there will be found fome practical diftinctions in the fymptoms, and remarks on the remedies, that are peculiar to ourfelves. On other difeafes it has been our wifh to add whatever our own practice has felected as ufeful, and, throughout the whole, we have ftrictly adhered to the duty of a faithful hiftorian.

Situated as I am at prefent, it is incumbent on me to confider every thing that is allied with my fubject; and my labours will be often found directed to objects entirely overlooked by my predeceffors. It was my fortune, as a medical man, to be introduced to the Navy early; fince which I have paffed through the gradations of mate, furgeon, and phyfician. To alleviate the miferies of human nature, it is neceffary to probe them to the bottom, and trace them to their fource; but to relieve effectually the diftreffes of a particular clafs of men, as the Britifh feamen, we muft affociate with the character, and keep aloof from none of their frailties. If this has been done with any fuccefs, the indulgent reader will forgive many errors, otherwife reprehenfible, in thefe pages. The whole were compiled from notes taken as opportunity offered, and with little premeditation. The ftudies of a naval phyfician are fomething like

. the

the foldiers prayers; they muſt be laid, " *when and where he can.*"

With reſpect to the arrangement: Two Diſcourſes are premiſed; in the Firſt I have made ſome obſervations on recent changes in the medical department, and what may be ſtill deemed objects of reformation. In the Second Diſcourſe, I have mentioned ſuch deſirable alterations as are connected with the preſervation of health, and the ſervice at large. A general abſtract is given of the health of the fleet, in one view; and then, a ſhort eſſay on the diſeaſes: the body of the work is ſo arranged, that future experience may be infuſed into the reſpective ſubject, or added in another volume, ſhould the war go on. This would particularly embrace the province of ſurgery. Some operations have been performed by our ſurgeons, with a degree of ſucceſs and elegance, that have been ſurpaſſed in no hoſpital whatever. I allude to two amputations at the ſhoulder joint, by Mr. HOUSEAL, of the Melampus; and one, under the moſt hopeleſs circumſtances, in a French officer, on board the Niger, by Mr. BURD; and theſe are a few out of a number that deſerve to be known.

It may now be aſked, whether or not this work might appear with more advantage at ſome future period, as has been cuſtomary with ſubjects collected during war. But very powerful

reaſons

reasons induce me to publish it at this time.
A practical subject can gain nothing by being
kept; on the contrary, it may lose much: facts
ought to be recorded as they first strike us, and
not as they may do at a long interval. It is true,
that more elegant language, and more correctness,
might have been employed, by protracting the
publication; but an unvarnished tale is preferable
to both. The body of the work might also have
been swelled, by contrasting my own opinions
with those of former writers; but that would
not have added to the value of the original. We
have, besides, been engaged in a department of
public service, where a progressive plan of improvement has taken place, from our suggestions,
under the auspices of a Board of Admiralty, whose
generosity has *answered every application* made to
them on the subject of health. If these applications have been seasonably made, and judiciously
administered when complied with, they will give
additional weight to our authority and advice in
recommending further alterations, and a moment
ought not to be lost in making them known.
There is, moreover, a responsibility attached to
the duty of physician to a large fleet, that I
should never wish to lose sight of. But the medical profession, as subservient to the happiness of
mankind, has claims upon my labours, that I
hasten to absolve. Such mistakes as become visible
here

here will be *beacons* for future adventurers; and if any thing is difcovered worthy of imitation, it ought to be followed.

It will be grateful to every lover of his country to read fo many fine traits of the naval character, which, as the Hiftorian of Health, have fallen to my lot to record. On the whole, it has been providential on my part, that my profeffional ftudies have been affociated with the fervices of a fleet, that have raifed the naval force of this country to the higheft pitch of glory. To the illuftrious Commander at its head, and other officers, the public has now frefh acknowledgments to make, for acts of kindnefs and condefcenfion to the health of the people, that have never been a fecondary confideration in our fhips.

It remains for me to make an apology, in this part, to Rear-admiral Sir ROGER CURTIS, Bart. for the liberty I have taken, in publifhing his Narrative of an infectious fever, on board the Brun.-wick, as lately printed, for the ufe of his friends, more efpecially as I never hinted this, when confulting him on other fubjects. My readers will confider themfelves highly obliged, and agree with me, that it merits a better fate than the circulation of a day, and could no where appear with more propriety, than in the hiftory of a fleet, whofe fick owe its author fingular obligations.

On

On a future occaſion I purpoſe to add a more accurate detail of the deaths, and the people ſent to hoſpitals; but this is a buſineſs for which I have no leiſure at preſent. From what is told here, may be collected a leſſon uſeful to poſterity; as it ſhews with what a ſmall expence of human beings a naval war may be carried on, when compared with the fate of the army on the continent and elſewhere.

I muſt now in the name of the Fleet, offer my grateful thanks to CHARLES DODS, Eſq; ſurgeon at the Royal Hoſpital, Haſlar, for the kind and candid ſupport he has uniformly given to the meaſures which have been deviſed by me to meliorate the condition of our ſick, and the general ſtate of the hoſpitals.

Similar acknowledgements are alſo due to Capt. JAMES STEVENSON, and the officers of the Charon Hoſpital Ship, for their attention to the ſick of the ſquadron under the command of Rear-admiral HARVEY, during the ſevere weather in October and November 1795, at Quiberon.

MEDICINA

MEDICINA NAUTICA.

DISCOURSE I.

MEDICINE, confidered as an inftitution for relieving and alleviating the diftreffes of mankind, is fo ingrafted on the cuftoms of civilized nations, that every well-ordered government has admitted its utility, and extended bountics to its profeffors. As a handmaid to the art of war, it has been infufed into our naval and military fervices; and to each a medical eftablifhment has been deemed a moft neceffary appendage. The neceffity of medical affiftance has been particularly obferved in modern wars. The colonies in Afia, Africa, and America, which belong to the powers of Europe, have, with their riches and commerce, brought a train of difeafes peculiar to their climate, and fatal to the conftitutions of northern nations, and which leave us to doubt whether or not we ought to regard thefe acquifitions as beneficial to fociety. The contefts, which have arifen between one country and another, and the jealoufy occafioned from the commercial confequence of

thefe

thefe poffeffions, have afforded a gloomy detail for the medical regifter during the laft century; but thefe are evils that have prevailed in all ages of the world: the fcenes that are now acting in Europe do not offer a more confolatory profpect to philanthropy. It is, therefore, the *tafk* of the medical profeffion, in the conflict, to ftep between fevere neceffity and the human fpecies; to wreft the innocent victim from difeafe; to calculate unwholefome feafons; to detect and deftroy contagions; to explore climates, fituation, foil, water, and air; and laftly, when the efforts of fkill can neither preferve or reftore health, they may tend in a great meafure to foften affliction, and "*fmoothe the avenue of death!*"

What has come to us, concerning the military operations of the ancients, is fo clouded with fuperftition and fable, that nothing is to be learned from it. Sea difeafes, in the ages of Greece and Rome, were altogether unknown, as their navigation was neceffarily coaft-ways, from the compafs being unknown. It was, therefore, the extent of navigation and commerce, that gave birth to a new train of diftempers, fome of which had been little known, if at all, to the ancient world. We muft, on that account, come nearer our own times to reap much information, and Dr. Lind may be juftly ftyled the father of nautical medicine. To his works have fucceeded thofe of

Robertfon

Robertſon and Blane; it may alſo be proper to mention with them, Dr. Clarke, who was a ſurgeon of an Indiaman, and wrote on the diſeaſes, of ſeamen.

Although the cultivation of medical ſcience, with all its auxiliary branches, has been much encouraged of late, yet our medical inſtitutions in the navy are far from having acquired improvement adequate to the object in view. This is partly to be attributed to the ſmall emolument, which any naval appointment holds out to a phyſician or ſurgeon; and partly to the want of a board of ſcience, to appreciate abilities, and reward induſtry in a profeſſion, that fundamentally differs from every other form of ſervice, and the knowledge of which is almoſt confined to its own members. To remedy theſe evils, a liberal and diſcerning Board of Admiralty have changed the commiſſioners for ſick and wounded, into a

MEDICAL BOARD.*

We are not yet informed what new privileges will be granted to this Board; but we muſt expect them to be of a very liberal nature, if they

are

* It may, perhaps, be acceptable to the medical reader of ſome future day, to know the names of the lords commiſſioners of admiralty, who have thus far contributed to the

general

are to extend the full advantages of the profeffion, to his Majefty's naval fervice. They will of courfe be invefted with power to act, in all cafes of emergency, with promptitude and decifion, as becomes the minifters of health. The progrefs of difeafe will not wait the return of poft, nor can forms of office ftay the ftroke of death. The appointment of furgeons has already been affigned to the phyficians, and with this it is fuppofed that new regulations concerning the examinations will alfo take place.

On a fubject, where my own opinions have contributed fomething to effect improvements, and where, for a length of time. I ftood alone, I

may

---

general welfare of the navy, and the medical department in particular.

JOHN EARL SPENCER, Firft Lord.

LORD ARDEN,   SIR PH. STEPHENS, Bart.
C. SMALL PYBUS, Efq.   J. GAMBIER, Efq. R. A.
LORD H. SEYMOUR, R. Adm.   WM. YOUNG, Efq. R. A.
                   EVAN NEPEAN, Efq. } Secretaries.
                   WM. MARSDEN, Efq. }

MEDICAL BOARD.

DR. ROBERT BLAIR, } Phyficians.
DR. GILBERT BLANE, }
SIR WM. GIBBONS, Bart. Civil Commiffioner.
JOS. STEWART, Efq. Secretary.

may now be allowed to add fome obfervations, as it will fcarcely be difputed, that my field for remarks has been extended, and my fources of information equal, if not fuperior, to any other perfon whatever.

In the little tract, which I publifhed feven years ago, on the medical department of the navy, I took up the fubject in a new manner, and confined the difcuffion of it to principles of fcience entirely. A longer experience, more extenfive acquaintance, and the general approbation of officers, induced me to keep the bufinefs ftill in view, in hopes that fome favourable opportunity might offer for a more perfect inveftigation. In a tafk of this kind, it cannot be expected that our actions will efcape either comment or calumny: one man fufpects that his indolence is impeached, another dreads the expofure of his ignorance, and a third may be afraid of fomething worfe than either. My promotion to Phyfician of the Fleet, has neither relaxed my endeavours, or made me lefs zealous to fee it accomplifhed, although I may be fuppofed not quite fo much perfonally interefted in the iffue as formerly. On the contrary, a large majority of the furgeons looked upon this appointment as a fortunate circumftance for their caufe: if they fhould be difappointed in the end, they muft accept the *will for the deed*.

It

It has been the good fortune of the medical profeſſion in Great Britain and Ireland, to enjoy emoluments far beyond what is to be acquired in any other country in Europe. In the metropolis, ſome of its members are now in the receipt of five thouſand pounds and upwards *per annum!* This liberal ſupport and reward of acknowledged worth and ſuperior abilities, may be juſtly attributed to wealth generally diffuſed among different claſſes of ſociety, and one of the happy conſequences of a free conſtitution and government. Many of the Phyſicians in London, and other great towns of England, began the practice of medicine in the army. Pringle, lately at the head of the Royal Society, was an army phyſician; but there are few of the navy liſt, that have been ſo fortunate in their profeſſional career. This can be imputed to nothing elſe, but different habits acquired by different modes of life; for the advantages of the early education are found to be much the ſame among navy and army ſurgeons. But the army ſurgeon has many advantages, in other reſpects, that can never fall to the lot of any perſon living in a ſhip: theſe are chiefly derived from being quartered in great towns, the intercourſe with poliſhed ſociety, and the gay manner of life peculiar to the army. Men attached to ſtudy and obſervation have it alſo more in their power to follow the bent of their inclination;

nation; they are not cut off from information to
be obtained from books; and their situation
affords them opportunities to cultivate acquaintance with literary characters, and the general progrefs of medicine. The very contrary
to all this is the condition of a navy furgeon; and
the fea-life affuredly begets a difpofition of mind,
that unfits him for the exercife of his profeffion
in private practice. He is, therefore, in the decline of his days, frequently left in a ftate of pre-
carious dependence : his naval fervitude, in the
prime of youth, had prevented him from making
friends and forming connections, that would have
been favourable to his future profpects in medical
rank and reputation; and his half pay is fo fmall,
that it only helps him to draw a comparifon,
painful to reflection, when he beholds the affluence of fome old fchool-fellow, who never
needed to encounter the toils and dangers of the
ocean, to earn his *otium cum dignitate.*

There are, moreover, a number of lucrative
appointments in the medical department of the
army, to which the furgeon has to look up, and
to which there is nothing equivalent in the navy.
The extent of the army ftaff at this moment,
abundantly fhows a far fuperior encouragement;
and the regimental furgeon has juft been allowed
a half pay of *five fhillings per diem.* When we
draw the comparifon between the one fervice and
the

the other, it is very natural to mark an obvious diftinction. We may not be confidered as their equals in medical knowledge and abilities; but in a faithful difcharge of our duty, to the beft of our power, we cannot yield up our claim to an equal indulgence from the bounty of the public.

When half pay was firft granted to the navy furgeons, it probably included the whole lift at the time, as the number of fhips was fmall. But as our navy increafed, the encouragement to the medical gentlemen does not feem to have been proportioned, and progreffive; for till lately one hundred and twenty-five, out of five hundred, received half pay. The victory of the 1ft of June, that will be memorable on many accounts, was an occafion where the abilities of the furgeons were generally noticed, and called for much exertion on their parts. It appeared alfo to give frefh fup-port, and a fit feafon, for renewing their claims for a further extent of the bounty of government. It became my official duty to mention to the Ad-miral and Captain of the fleet, the meritorious conduct of thefe gentlemen; and their fituation had frequently been the fubject of previous con-verfation with both. I was made to underftand, that the commander in chief would moft cordially fupport any application the furgeons might make to the lords commiffioners of admiralty, as a tefti-mony of his approbation of their recent fervices.

This

This welcome intelligence being communicated to Mr. Peter Smith, Mr. Stephenson, and Mr. Glegg, a general meeting of the furgeons took place, and a fhort, but impreffive petition was tranfmitted to the fecretary of the admiralty, to be laid before their lordfhips.* From this time the lords commiffioners were pleafed to order the neceffary·information to be laid before them; but before definitive arrangements could be made out, very confiderable changes took place at the admiralty.

When the fleet was called out in the winter, there were upwards of forty vacancies for mates throughout the fhips, which afforded the admiral another opportunity to requeft their lordfhips attention to the bufinefs: the board was juft formed under Earl Spencer. Their lordfhips were, therefore, pleafed to decree an increafe of half pay, and other encouragements, as they now ftand; and they are great indeed compared with the former eftablifhment.

The pay of the individual, and the intereft of the public fervice, in this department, are fo infeparably connected, that it becomes neceffary to

---

* At this time their lordfhips were pleafed to increafe the pay of furgeons mates 1l. per month: if the firft mate was in poffeffion of a fet of inftruments, he was decreed to receive 5l. per month.

C

in

combine them in this difcuffion: improvements in one, can, only take place with an increafe of the other. When the late additions were made, it was obferved, that a number of furgeons came forward, who had declined employment at the beginning of the war. I confider it as much a part of my duty. to record, thefe circumftances, as any other tranfaction. But a.total change in the mode of payment muft be the firft ftep towards a fcientific fyftem. It muft be a fixed fum, whether by the day, or by the month, and the half pay modified by it. This was the object I had in view in my former work, and what improvements have been adopted lately, are but fo many advances towards it. There will be found in thefe pages ample proofs, how much,. on certain occafions, the defence of this country may depend on a vigilant and active medical practice; which in their proper place will plead a fufficient excufe for the warmth with which I muft continue to recommend further alterations.

It requires no eloquence of mine, or that of any other perfon, to awaken the fenfibility of an Admiralty Board, that has already looked fo kindly to the fick-bed of the failor: but for many reafons it is to be wifhed, that all improvements, which by the directions of their Lordfhips are to defcend from the Medical Board, fhould be brought to trial before the conclufion of the war.

6

I lament

I lament that an hour fhould pafs before the completion of a fcheme that cannot fail to give frefh refources to our naval department.

The fupply of medicines claims immediate correction from the Medical Board. It is at prefent interwoven with the pay, and the fource of multifarious abufes. Thefe, like fome others, can only be prevented by changing the mode of payment, and allowing the medicines at the expence of government. The full pay of the furgeon has never been the caufe of complaint, unlefs in very fickly fhips; and in fuch fituations it is liable to engrofs the whole of the emoluments. But it is a liberal half pay, that is to anfwer the magnitude of the object, utility to the fervice, and fupport to the individual; and to make it equal to that of the army furgeon, could only be doing away an invidious diftinction.

I am ftill of opinion, that the fupply of medicines to his Majefty's fhips, can only be effectually done at government expence, and under the controul of the Medical Board. Some time ago, when confulted on this bufinefs, I fketched out a plan, down to the minuteft forms of office. The outlines were, to erect Difpenfaries at the dockyards of Chatham, Portfmouth and Plymouth, to be fuperintended by a navy furgeon, with clerks, porters, &c. Branches of thefe inftitutions, for the fupply of fhips abroad, fhould be eftablifhed

at

at Kinfale, Gibraltar, Madras, Calcutta, Antigua, Jamaica and Halifax. A fum of money, not exceeding 20 *l.* fhould be allowed the furgeon, on his firft appointment, to furnifh inftruments, which he is to repair and recruit at his own expence afterwards: thefe are to be annually furveyed at the Difpenfary, for which he will receive a certificate; and when unemployed, they are to be depofited in one of the Difpenfaries. Medicine chefts for different rates will be appointed by the Board, and alterations made in their contents, to correfpond with the nature of fervice and ftation of the fhip. Occafional fupplies fhall be made, as in other branches of fervice. He fhall make oath, once a year, on pafling accounts to receive pay, that thefe articles of medicine have been duly and faithfully adminiftered on board his Majefty's fhip. From thefe Difpenfaries all the furgeon's neceffaries fhould be iffued under fimilar forms. It appears to me, that an inftitution of this kind, to anfwer all the purpofes of fervice, might be brought into actual practice in a few weeks.

After the return of the fleet to port, in June 1794, we accomplifhed a very defirable change in the neceffaries, by throwing out fome ufelefs articles, that quickly fpoiled, and others that engroffed a large part of the money, though of little ufe. It was of great confequence to procure more tea, which of all articles in diet is moft relifhed

by

by our fick : the whole alterations were limited to the fum allowed for the old form. A feventy-four, for one month, receives,

NEW FORM.*			OLD FORM.		
		yds.			*lb.*
Finer new linen	—	12	Lump Sugar	—	48
Welch linen (bandages)		8	Tea	—	1
			Currants	—	20
		lb.	Rice	—	18
Tea	—	8	Barley	—	18
Cocoa, or Coffee	—	12	Sago	—	10
Sago	—	8	Almonds	—	$1\frac{1}{2}$
Rice	—	16	Tamarinds	—	3
Barley	—	32	Garlick	—	4
Fine foft fugar	—	64			
Ginger	—	$0\frac{1}{4}$			*oz.*
			Shallots	—	8
			Mace	—	2
Saucepans, ftrong	—	4	Cinnamon	—	4
Canifters	—	2	Nutmegs	—	2
Boxes	—	1			
					yds.
			New linen	—	12
			Saucepans	—	4
			Boxes	—	1
			Canifters	—	2

* Thefe articles being now fufficient for the purpofes of fervice, it only remains to allow an addition to the fum, that the quality of each might be improved.

When an increafe of falary was ordered to the Medical Board, we are led to regret, that Govern-

ment

ment did not at once make it equal to the Army
Board. The phyficians, who are to fuperintend
the health of the Britifh navy, fhould be furpaffed
by none in fplendour of eftablifhment; and it
would have put it more in their power to enforce
fome regulations, that are effentially connected
with its welfare, and to which all eyes are, at this
moment, fixed. I would have their lordfhips to
proceed with this work of benevolence, and not
to confine the limits of it, by comparing what is
done with what it was before, that the fpirit of the
undertaking may never be damped by any ill-
timed parfimony. It can be deemed no breach
of national œconomy, to reward the fervices of
eminent and learned characters, in fuch a manner
that the public may enjoy the full benefit of their
abilities. It is not in Board hours alone, that
thefe talents will be ufefully employed: their
private, as well as official correfpondence, muft
pervade every corner of the department. Such is
the nature of medical fcience, that as it is conftantly acquiring new acceffions of knowledge,
from new facts and obfervations, thefe phyficians
will ferve as a *focus*, where every fcattered ray is to
be condenfed, and where every new idea will
again diverge, for the information of the whole.
Our prefent labours are a humble epitome of this
manner of collecting facts, and an imitation of
what Dr. Blane was the inftitutor, in a former war.

When

When the younger members of the profeffion are aware of a fcientific Board to watch their labours, it will tend to ftimulate genius and induftry, and will occafionally draw forth valuable talents, that might be left to ruft in obfcurity, or remain neglected, to the lofs of themfelves and the country. By thefe means only will Phyficians at the Board be able to confer that diftinction on merit, which muft always be expected from their hands; and which I conceive to be a facred part of the truft confided to them, as the directors of medical fcience in an extenfive public fervice.

HOSPITALS.

A word at parting : Much of my profeffional labours having been directed to the reformation and improvement of the Royal Hofpitals, it remains for me to add fomething here, as peace does not feem very diftant ; for I fhould be forry to with-hold a fingle idea that can affift the completion of a work, to which the interefts of the navy are fo nearly allied.

When I was honoured with the appointment of Phyfician to the Fleet, I deemed it my duty to fubmit to fome officers high in command, what I conceived to be deficiencies in thefe inftitutions, and whom, I knew, would intereft themfelves in

C 4 the

the bufinefs. The Commander in Chief, as the
proper mode for official inquiry, ordered Admirals
Caldwell and Gardner, with Captains Domet and
Nichols, to furvey Haflar Hofpital, and report
to him their remarks. This duty was conducted
by thefe officers with great patience and attention,
and fome very material changes have taken place.
In an undertaking of this kind, but little forefight might affure me that I was foon to meet oppofition to my projects. Men, who were 'my
feniors in years, and fuperiors in knowledge, were
quickly ruffled at the idea of change, in a fituation
where all had been thought perfect, and where the
fmalleft ftep towards reformation had never been
judged expedient. Thefe inftitutions were particularly deficient in naval officers to keep the
feamen under difcipline and command; in the pay
of phyficians and furgeons, with other officers;
in the number of phyficians and furgeons; in
buildings to lodge the officers; in the internal
œconomy of the hofpital, fuch as adminiftration
of medicines, diet for the fick, wafhing of the
cloathes, &c. My remarks on thefe fubjects were
afterwards printed, with curfory arguments, for
the information of public boards and officers,
without a view to publication. Here, as with
the navy furgeons, I contend, that the want of
improvement has been owing to the want of encouragement; and till it is made equal to the
duty

duty of the truſt, the ſervice muſt occaſionally ſuffer: it is becauſe I value the abilities of theſe gentlemen as much as I ought to do, that I wiſh to ſee them amply provided for. Thoſe who know me beſt, will teſtify how little any thing perſonal has influenced my criticiſm; and thoſe who wiſh beſt to the navy, muſt regret that any member of a public department ſhould be obliged to receive part of his ſupport from private practice,

Anxious as I always muſt be for the welfare of the navy, I cannot help feeling, when I look to the pay of phyſicians on the army ſtaff, and compare it with the naval hoſpitals. To theſe ſtations, the induſtrious ſurgeon of a ſhip, who has to paſs a youth of turmoil and care, ſhould be taught to look up, as an incentive to emu- lation, and premium of ſervice. Men who can fill theſe appointments with advantage to the navy, and reputation to themſelves, a diſcerning Medical Board will always be able to ſelect. In ſhort, this Board has the happineſs of thouſands in their hands; by directing the ſtreams of the ſcience to flow from themſelves to every indi- vidual, to give health to the ſick and comfort to the diſtreſſed. Their precepts will be revered, and their example imitated: what is no inferior conſideration in the medical character, they will alſo extend the kind offices of humanity to the
ſick

fick bed. .It is the lot of the failor and foldier to languiſh under affliction and difeafe far from the cheering fupport and watchful attendance of friends and relations * ; and hence a charge of another kind devolves on their phyfician, that nurfes and others may be tender and affiduous in their refpective duties,

Would not a medical library be a valuable appendage to thefe hofpitals ?

Might not the ground round Haflar be advantageoufly laid out in gardens and orchards for the ufe of the fick ? How grateful is a diſh of fallad after a long cruize ? How delicious an apple, a pear, or a plumb, after a long ficknefs on board ? Speak you, who, like me, have had three narrow efcapes from death, and a long confinement, with a tardy convalefcence, at fea ! To thefe I would add a pigeon houfe, and poultry farm ?

Amidſt other deficiences, I think a fuit of baths one of the greateſt. There are many difeafes peculiar to naval officers and feamen, where they would be of infinite fervice; and it is furprifing they fhould have been fo long neglected, The fituation of both Haflar and Plymouth Hofpitals are convenient for their conſtruction : they

---

* Heu, terra ignota, canibus data præda Latinis,
,   Alitibufque jaces ! nec te tua funera mater
    Produxi, preffive oculos, aut vulnera lavi.

ÆNEID.

ought

ought to be made for different temperatures; and joined to a magnificent national charity, the baths of Haflar would become famous as thofe of Baiæ, in the days of ancient Rome.*

We may with thefe improvements hope to fee an apparatus for the exhibition of *factitious airs*: for furely an inftitution like Haflar Hofpital ought to be the firft to introduce difcoveries into practice, that promife and have already effected cures in difeafes, for which the *Materia Medica* had been ranfacked in vain.

" Jam nova progenies cælo demittitur alto."

The firft part of my attendance at Haflar Hofpital was marked by a defire to correct any forms, that might not be confiftent with my own ideas of medical practice; I therefore began with the following regulations, which would have been altered and improved, as a longer acquaintance

* Tubs were employed for this purpofe, while I belonged to Haflar, in the form of thofe ufed in flop fhips, for purifying new-raifed men; but the feamen had fuch a diflike to them, that it was found impracticable to get a rheumatic patient to bathe, becaufe they reminded them of *fcrubbing*, by way of punifh-ment, on board. Inftruments of this kind degrade a public charity; a failor under difeafe ought to be bathed like a gentleman.

with

with the duties of my ftation might fuggeft. The impartial reader will obferve that, in any changes which I have prefcribed in thefe inftitutions, no reftraint has ever been offered to others, that was not adhered to by myfelf. In an hofpital like Haflar, I think the practice of the medical profeffion might be carried to a higher degree of perfection, than in any other in Europe; and the pure fpirit of the fcience made fubfervient to the treatment of the objects it is intended to relieve, beyond what can be done any where elfe. If I am thought to be fingular, or Utopian, in this opinion, I have no objection to ftand alone.

DIRECTIONS,

For the Vifiting Apothecary, and Affiftant Difpenfers, in the wards of the North Wing of the Royal Hofpital at Haflar.

I.

They fhall fee that the general regulations of the hofpital, as put upon the ward doors, are duly attended to by the patients, nurfes, &c.

II.

The Difpenfers fhall go through the different wards, allotted to their fuperintendance, every morning: from March to September, at eight o'clock;

o'clock; and from September to March at nine o'clock, to fee that the wards are all difpofed in clean order and in due time, viz.

1. The patients in the Recovery Wards to be out of bed.

2. The beds to be properly made up.

3. The wards to be cleanly fwept afterwards.

4. All pots and offenfive matters to be carried out and purified.

5. The doors and windows to be opened for the purpofes of ventilation.—This to be done, by the fafh of one window being put down, while that of the next is thrown up; and fo with the others.

6. They are to enquire of each nurfe, who has had the watch, whether any thing particular has occurred in the night, and to act accordingly, or report it to the Phyfician.

III.

Every patient is to be vifited immediately on coming to his ward, and what may be deemed neceffary, prefcribed.—The Gentlemen are defired to be attentive, in examining the duty of the labourers employed in cleaning and wafhing the people; to return 'em if not fufficiently cleaned, and the neglect to be reported to the Phyfician.

IV. Patients

IV.

Patients with contagious fevers, fmall pox, and meafles, are to be ftrictly feparated from others, and no vifitors are to be admitted into thefe wards. Thofe labouring under chronic complaints of the fame nature, are to be put together, as nearly as circumftances will admit of.

V.

Cleanlinefs being one of the greateft requifites of an hofpital, in all their vifits this is to be particularly watched, and the nurfes are to be charged with it ; viz.

1. The men are to be kept clean in their perfons, by frequent changes of bodylinen, &c.
2. By wearing their hair fhort.
3. By wafhing themfelves every morning.
4. The nurfes are to wafh thofe who are unable to do it themfelves.
5. Their beards are to be regularly fhaved.
6. The bed-linen is to be duly fhifted, and the beds as often as occafion may require.
7. All clofe-ftools and bed-pans, &c. to be emptied, and wafhed immediately after ufe.

8. In

8. In non-compliance with thefe rules, the nurfe or patient to be reported.

VI.

As perfonal flovenlinefs is difagreeable in men, fo it is difgufting in women : when obferved, it is to be reported to the Matron.—This rule extends to keeping clean and neat,

1. Their cabbins,
2. Floors and ftairs,
3. Water-clofets, &c.

VII.

The attendance of the fick in bed is to be particularly watched ; that they have their medicines duly; that they are regularly ferved with diet and drink ; and that no nurfe be permitted to treat them harfhly or unfeelingly.

VIII.

As it becomes the duty of every medical attendant to fuperfede the ufe of medicine by diet, when it can be done, fo the appetites and cravings of the patients, for any particular kind of food, are to be regarded ; and as far as the regulations of the hofpital admit, are to be complied with.— In cafes of debility, and want of the ufual defire for animal food, wine will be grateful, either by itfelf, in the drink, or with fago, rice, panado, &c.

IX. The

IX.

At the hour that the medicines are received into the wards from the Dispensary, the assistant dispensers shall attend in their respective wards, to be certain that each man has received his medicines, as prescribed on the ticket, that it may on no account be left to the nurses, as hitherto.— At this time they are also to be assured, that every patient is in his proper ward, and to permit none to leave their wards, afterwards, for the night.

X.

The garret wards being solely reserved for men about to be discharged, no patient is to be moved there that takes medicine; but in cases where this may be again necessary, the patient is to be moved into another ward, more under the inspection of the Physician.—Fever patients, with surgical complaints at the same time, are to be sent to Lobby 62, to be visited by both Physician and Surgeon.

XI.

The Physician will begin his morning visit at No. 81, and finish with the small pox wards, No. 103 and 104. The visiting apothecary and dispensers, are desired to begin and finish their rounds in the same manner.

XII. The

XII.

The Phyfician will prefcribe to all thofe cafes which more particularly require his attendance; but all changes of importance in the ftate of any patient, are to be duly reported to him.—His prefcriptions, for the fake of diftinction, will be written with *red ink*; and they are not to be altered without a material change of fymptoms, or in the recovery of the fick man.—The vifiting apothecary fhall fee thefe patients as often as the nature of their complaints may require, and attend to the exhibition of their medicines.

XIII.

The Difpenfers are defired, at *unftated times* of the day, to walk through certain wards, in order the more readily to detect all deviations from the modes of regulation and difcipline it is wifhed to inculcate: many opportunities for difobedience may occur, whether from diforderly nurfes, or by men who have *feigned* complaints for the purpofe of being invalided.—The affiftant difpenfers in the wards fhall perform this duty in weekly rotation.

XIV.

As it appears abfolutely neceffary, for the benefit of his Majefty's fervice, that a regular

D   fyftem

fyftem of duty, among the medical attendants, fhould be eftablifhed, whether for punctual vifits to the fick, or to make themfelves fufficiently acquainted with the characters and difpofitions of men who practife deception, the gentlemen are required to give the neceffary orders to the nurfes and others; and to affure them that nothing can be forgiven, that infringes againft thefe rules : the Phyfician therefore expects that they fhall be ftrictly adhered to in all departments.

T. TROTTER, Phyfician.

Royal Hofpital,
    Feb. 4th, 1793.

MEDICINA NAUTICA.

———

DISCOURSE II.

THE character of a British Seaman exhibits so many striking singularities, that blend themselves so much with all his habits, that a thorough acquaintance with them becomes neceffary to both officer and phyfician, in their refpective ftations. Thefe peculiarities are the offspring of a fea-life, from the little communication it affords with the common manners of fociety. The love of adventure and enterprize, that fo foon difcovers itfelf in an active boy, feems to prompt the firft inclination for fea; a longing curiofity keeps it alive, and nothing but a voyage will at laft fatisfy the youthful Argonaut; to which the parent confents, in the hope that a life of danger and toil will foon ficken the unexperienced failor, and make him wifh to live at home. This, however, feldom happens, and the firft cruize or voyage cafts the die for a future fea life to the young adventurer. It is fomewhat remarkable, that boys in inland towns fhould fo

often

often fhow this early defire of going to fea. I
have, however, feen it difcover itfelf there in a
very romantic manner, and terminate in an
elopement purpofely to embark: among boys of
this defcription, the hiftory of a broken failor is
accounted the fineft piece of eloquence; and
whenever he appears, the narration of his voyages,
battles, and fhipwrecks, are liftened to with rapture.
The voyages of Drake and Anfon round the
world, are famous in this way, and eagerly read
by fchool boys; but Robinfon Crufoe has made
more profelytes to thefe kinds of adventures, than
all other mariners: his ftory, from firft to laft, is
fo full of incident; in all his difficulties he fhows
fo much courage, addrefs, and ingenuity, that the
young reader fancies himfelf the difcoverer of
fome great kingdom, and his imagination wanders
for ever in queft of an ifland. Even the Englifh
newfpapers, now fo generally circulated, have a
wonderful effect in fpreading this enthufiafm for a
fea-life: the number of well-fought actions, between fingle fhips, during the prefent war, will
cherifh it, and fhape the fortune of fucceeding
warriors; while the fublime manœuvre of piercing
the French line, by Earl Howe, will be equally
appealed to, at fome future day, by the hiftorian
and fchool-boy.

In a country, like this, that owes her fecurity
to a naval force, we fee a victory at fea celebrated
                                            above

above all others; it rouses the *amor patriæ* to the highest pitch of enthusiasm, and reminds a free people of their independence ; because nature has decreed that this is our element. The names of our great admirals are therefore revered as so many tutelary deities of 'our island—Hawke, Rodney, and Howe—and the heroes of the Granicus and Rubicon shrink into insignificance, when compared with those of the 12th of April, and the 1st of June.—Hence, from peculiar causes, the naval spirit of Great Britain, descends, as it were, in hereditary succession.

That courage which distinguishes our seamen, though in some degree, inherent in their natural constitutions, yet it is increased by their habits of life, and by associating with men, who are familiarized to danger, and who, from national prowess, consider themselves at sea, as rulers by birth-right. By these means, in all actions, there is a general impulse among the crew of an English man of war, either to grapple the enemy, or lay him close aboard : French men shudder at this attempt; and whenever it has been boldly executed on our part, they run from their quarters, and are never to be rallied afterwards. Nor does this courage ever forsake them; we have seen them cheering their shipmates, and answering the shouts of the enemy, under the most dreadful wounds, till, from loss of blood, they expired.

D 3   It

It is only men of such description, that could undergo the fatigues and perils of a sea life ; and there seems a necessity for being inured to it, from an early age The mind, by custom and example, is thus trained to brave the fury of the elements, in their different forms, with a degree of contempt, at danger and death, that is to be met with no where else, and which has become proverbial. Excluded, by the employment which they have chosen, from all society, but people of similar dispositions, the deficiencies of education are not felt, and information on general affairs is seldom courted.   Their pride consists in being reputed a thorough bred seaman; and they look upon all landmen, as beings of inferior order. This is marked, in a singular manner, by applying the language of seamanship to every transaction of life, and sometimes with a pedantic ostentation.   Having little intercourse with the world, they are easily defrauded, and dupes to the deceitful, wherever they go: their money is lavished with the most thoughtless profusion ; fine cloathes for his girl, a silver watch, and silver buckles for himself, are often the sole return for years of labour and hardship. When his officer happens to refuse him leave to go on shore, his purse is, sometimes, with the coldest indifference consigned to the deep, that it may no longer remind him of pleasures he cannot command. With minds uncultivated and uninformed, they are equally credulous and superstitious:

x

ſtitious: the appearance of the ſky, the flight of a bird, the ſight of particular fiſhes, ſailing on a certain day of the week, with other incidents, fill their heads with omens and diſaſters. The true-bred ſeaman, is ſeldom a profligate character; his vices, if he has any, rarely partake of premeditated villany, or turpitude of conduct; but rather origi-nate from want of reflection, and a narrow under-ſtanding. Hence he plays the rogue with an awk-ward grace, though the degree of cunning which he occaſionally practices towards his creditors be-ſpeaks art: but from them he has learned the way to over-reach; and it ought to be remembered, that they have a particular intereſt in emptying his pocket as quickly as poſſible; for his bargains with the world, are limited to his landlord and ſlop-ſeller. In his pleaſures he is coarſe, and in his perſon ſlovenly: he acquires no experience from paſt miſfortunes, and is heedleſs of futurity. His converſation, commonly, turns upon his own profeſſion, and his animadverſions are almoſt con-fined to a ſhip, her various properties, ſuch as ſailing, rigging, &c. yet the ſailor has a wit of his own, and he tranſlates all occurrences into his own phraſes: *cunns* a horſe when he rides; *heaves the lead* from the top of a ſtage coach, and *wings* * his enemy, when he ſhoots away his ſtun-ſail hal-liards. Thus his narrations are full of hyperboles,

* Captain M'bride's Letter to the Admiralty, Nov. 1781.

fimilies and comparifons : and if he finds he can
work upon the credulity of his hearers, he will
frequently outdo De Foe or Gulliver himfelf:

——— —— — —— even from his boyifh days
Till the very moment that you bade him tell it.
Wherein he will fpeak of moft difaftrous chances,
Of moving accidents by flood and field ;
Of hair-breadth 'fcapes, i' th' imminent deadly breach :
Of being taken by the infolent foe,
And fold to flavery ; of his redemption thence
Wherein of Antres vaft, and defarts wild,
Rough quarries, rocks, and hills whofe heads touch Heaven ;
And of the cannibals that each other eat,
The Anthropophagi, and men whofe heads
Do grow beneath their fhoulders.

                                SHAKESPEARE.

Some new traits are engrafted on the character,
by coming on board a man of war, and to be
traced to the cuftom of impreffing them.   This
is apt to beget a fulkinefs of difpofition, which is
gradually overcome, when he recollects that he
only refigns his own liberty for a feafon, to become
a champion for that of his country.   It, however,
often preferves a determination to watch every
opportunity for effecting his efcape : it is alfo the
fource of numerous deceptions, by making him
affume difeafes, to be an object for invaliding.
Hence he employs cauftics, to produce ulcers; in-
flates the urethra, to give the fcrotum the appear-
                                            ance

ance of hernia; and drinks a decoction of tobacco, to bring on emaciation, sickness at stomach, and quick pulse. Under trials of this nature, there is exercise for both patience and discernment on the part of the officer and surgeon; but there is rarely occasion for punishment. A well-regulated ship, soon reconciles all disaffection. This war has been singular for few desertions; and general punishments have scarcely been known in the Channel Fleet. His real diseases spring from causes peculiar to a sea life: laborious duty, change of climate, and inclement seasons, bring on premature age, and few of them live to be very old.

If such are the follies and vices of the sailor, his virtues are of the finest cast. In the hour of battle, he has never left his officer to fight alone; and it remains a solitary fact in the history of war. If, in his amours, he is fickle, it is because he has no settled home to fix domestic attachments: in his friendships he is warm, sincere, and untinctured with selfish views *. The "*heaviest of metals*," as

Sterne ·

* When a ship comes to action, it usually puts an end to all party quarrels among the officers and people; so ready are they to unite against the common foe of their country. It also tries the sincerity of friendships.

On board the Berwick, where I was surgeon's-mate, in the action with the Dutch fleet on the Dogger Bank, Aug. 5th, 1781, a sailor was seized with convulsions immediately on being

Sterne calls it, becomes light as a feather, in his hands, when he meets an old shipmate or ac-

ing informed that his messmate was killed. They had lived three years together: whenever the mess sat down to take victuals, this man's convulsions always recurred, and the seat that his friend used, always brought him to his mind, when he looked at it. It was a very affecting sight to the officers.

In the same battle, a young midshipman who was wounded on the forecastle, could not be found after the action, and was supposed to be thrown over-board. Another young gentleman, who lived with him in the gunner's mess, became inconsolable, for the loss of his friend; never was grief so exquisite.

> Ut stetit et frustra absentem respexit amicum:
> Euryale infelix, qua te regione reliqui?
> Quave sequar!———
> Tantum infelicem nimium delexit amicum.

It happened, however, that, stunned with a blow which he had received on the head, he creept into the bread-room, where he fell asleep, and did not wake for some hours. On coming out, he was perceived by his friend—they flew to one another; while the spectators wept at the interview: it could only be exceeded by the Roman Matron, who died with joy, on being informed, that her son survived the battle of Cannæ. These youths were twelve or thirteen years old: one of them had the honour of being signal officer, in the Queen Charlotte, on the first of June: they are both captains of the navy at this moment, with the hopes of their friends and country, fully realized in their professional accomplishments.— What a nice test of friendship!

> Fortunati ambo! si quid mea carmina possunt.
>   VIRGIL.

acquaint-

quaintance under diftrefs: his charity makes no
preliminary conditions to its object, but yields
to the faithful impulfe of an honeft heart. His
bounty is not .prefaced by a common, though affected harangue, of affuring his friend that he will
divide with him his laft guinea: he gives the
whole; requires no fecurity, and cheerfully returns to a laborious and hazardous employment,
for his own fupport. Was I ever to be reduced
to the utmoft poverty, I would fhun the cold
threfhold of fafhionable charity, to beg among
feamen; where my afflictions would never be in-
fulted, by being afked, through what follies or
misfortunes I had been reduced to penury.

Having faid fo much of the failor, it may be
expected that I fhould add fomething on the
officers; but I have affociated too long with the
character to be deemed an impartial delineator.
The country will learn their value, from recent
and matchlefs fervices : and fome of the gentler
virtues which adorn the naval profeffion, in watching the health and comfort of their people, will
receive abundant teftimony from the following
pages.

RAISING MEN FOR THE NAVY.

The experience collected on this fubject, during the prefent war, entirely correfponds with the
account

account of numerous evils, related on former occafions. They muft be inevitable, till a new fyftem is adopted. A country, that boafts fo juftly
of her *civil rights*, ought long ago to have refcued
from an involuntary engagement, a defcription of
people, to whom fhe owes her greatnefs in the *fcale*
of empire. I am afraid that men high in office,
have a very limited idea of the afflictions occafioned
by impreffing feamen. Inftead of calling it, a neceffary and politic meafure, for the fafety of the country, I pronounce it to be a moft fatal and impolitic practice.   It is the caufe of more deftruction
to the health and lives of our feamen, than all
other caufes put together ; and every nerve of invention ought to be ftrained, to put a fpeedy
and effectual check to it.

   We have at laft found an alternative for preff-
ing : the *requifition* of feamen and landmen for
the navy, which was made in the fpring of 1795,
by act of parliament, brought into the Houfe of
Commons by Mr. Pitt, has formed a precedent
that ought to be imitated on every future emergency. I am only forry that this act was not made
permanent, fo as to enable minifters to call upon
the counties and towns, whenever a levy of men
was found neceffary.   There is not an objection,
of any force, to be offered againft a repetition of
this kind : and had the officers, who regulated the
volunteers, been fomewhat more attentive in ex-
                                          amining

amining them, there remains no doubt, but it would have effected all that was wished for the good of the public service. At some places, very high bounties were given, even to forty guineas, which were the cause of much fraud and imposition : men utterly unfit for duty, but with no apparent disease, entered for the sake of this sum, and after being a few months or weeks on board, discovered their complaints to get invalided. It was particularly hard to press an able seaman, after such high bounties had been given to landmen ; for the king's bounty and these bear no comparison.

In the beginning of a war, if we suppose the peace establishment to be twenty thousand men ; sixty thousand more may be raised by requisition, in the like manner, and in the space of four months ; by which means, seventy sail of the line, with a proportion of smaller vessels, would be ready to strike a blow before any enemy could be prepared to face us.

The evils of impressing are manifold : a great number of our best seamen immediately disappear at the beginning of a war, and conceal themselves. It requires some time to get ships and tenders ready ; the people are crowded together ; they sleep on the decks ; they are without cloaths to shift themselves ; persons of all denominations are huddled together in a small room, and the first twelve months of a war afford a mournful task for

the

the medical regifter, in the fpreading of infection, and fickly crews. Hence a new commiffioned fleet of fhips can never be deemed an effective force at the early commencement of hoftilities.

In a country like this, where fo large a proportion of the people are employed in manufactories of different kinds, it is the firft effect of a war to throw many of them out of employment. The only refource is the navy and army. We muft fuppofe that men of this defcription, at leaft that are married and have children, leave a fituation where they have had enjoyments, and have to look forward to one where thefe blandifhments are to have no fhare; confequently, they muft feel thofe pangs of feparation at leaving a virtuous charge, which are natural to human beings : this affecting tale needs not the language of romance to find its way to feeling hearts ; but muft now and then be aggravated to the moft poignant diftrefs, by refigning wives and children to beggary and want, and a thoufand ills of which I can form no idea. Hence that dejection of fpirits that makes them the firft fubjects for the fcurvy, and the earlieft victims to contagious difeafes. Many a melancholy ftory is thus related to the medical attendant of a failor or foldier, and it begets a fympathy that interefts us the more in their recovery: under this fpecies of mental affliction we know that numbers perifh, without any apparent diforder.

Now

Now if thefe people were all levied by the re-quifition bill, their bounties would be fo confider-able, that a moiety could be left for the fupport of their connections, which I know was very gene-rally the cafe on the late occafion; and it tends to alleviate the pains of feparation. The poor land-man receives a bounty of twenty or thirty fhillings, which buys his firft jacket; he probably paffes through two receiving fhips, and three or four ten-ders, before he arrives at the fhip where he is to be ftationary. By this time frefh flops are wanted, and the firft year of his fervitude does not put a fhilling in his pocket. Very different is it with a requifition man; he can be trufted to march by land, and go to a king's port at once, without incurring difeafes from the paffage in a tender. He alfo confiders himfelf a volunteer, and feels nothing to deprefs his mind, or to prevent him from accommodating himfelf to the cuftoms of a new fituation. He is thus reconciled to the navy, thinks lefs of trying to defert; or, if he did, he runs a greater rifk of being taken up, as he may be known or heard of on the fpot where he en-tered.

It may be faid that high bounties were a heavy tax upon individuals: but if we were to calculate the vaft faving of human lives, and the money fpent in raifing men under other modes, it will be found the very quinteffence of œconomy. But, putting

putting that out of the queftion, what a trifle
ought it to be reputed, when it prevents an Englifhmen from being impreffed, and makes him a
volunteer in the fervice of the public. I muft,
therefore, congratulate the country on the efficacy
of Mr. Pitt's bill, and hope to fee it made permanent. My authority ought to give fome weight
to the repetition, for I have feen much of its good
effects, and have often witneffed the horrors of the
old fyftem.

D R E S S.

A general uniform for feamen has been mentioned in the valuable works of Lind and Blane,
and fupported by arguments fo conclufive, that
nothing can be offered againft them. Amidft fo
many improvements in the navy, we are furprifed
that fuch a one as this has never been brought into
univerfal practice. A uniform in all fituations
contributes fo much to perfonal delicacy and
cleanlinefs, that we are at a lofs to conceive how
our officers have neglected it fo long: it is the more
to be wondered at, as the moft punctilious attention is now paid to the cloathing of the people.
The Hon. Captain George Berkeley, when he
commanded the Magnificent, as a guard-fhip at
Portfmouth, had his men dreffed in a particular
way: they were eafily diftinguifhed from others,

and

and became proverbial for neatnefs of appearance,
and orderly behaviour when on fhore. Nay, fo
little have thefe ideas refined old habits in the
fervice, that the navy flops are really made in a
form that no failor, who has any tafte in drefs, will
put them on. But there are other arguments to
be ufed in favour of a general uniform. It appears to me, that defertion would be very much
prevented by it: at leaft, it would increafe the
difficulty of efcape, as difguife could not be fo
eafily affumed. The cloathes might be manu-
factured of a particular kind of cloth; and an act
of parliament paffed, enforcing the fame regulations
and penalties as are ufual in the army. The
uniform fhould confift of a blue jacket, with a
fleeve and cape of the fame, and lined with thin
white flannel: a waiftcoat of white cloth, trimmed
with blue tape: blue trowfers, or pantaloons, of
the fame cloth with the jacket, for winter; and
linen or cotton trowfers, either ftriped blue and
white, or all white, for fummer: check fhirt,
and black filk neckcloth. A button of metal,
or horn lefs liable to tarnifh, with the letters R. N.
upon it. The hat fmall and round, water proof,
with a narrow belt, on which fhould be printed the
name of the fhip; which could be conveniently
fhifted when a man is turned over to another fhip.
An outfide jacket, of a thicker texture, and flannel
waiftcoats, might be occafionally fupplied, as a de-

E   fence

fence from cold and rainy weather. Such a form
of drefs could not fail to be acceptable to the
feamen; and it would be highly pleafing in the
eyes of officers and others. The crews of different
fhips would be known by the name on the hat-
band, which would make them emulous to appear
clean and orderly: this again would increafe attachment to the fervice and its commanders, and
with thefe all the virtues of good difcipline.*

D I E T.

This is a department where great improvements
have been made of late years. The falted beef
and pork are excellent; and the bread, till the
high price of corn rendered a mixture neceffary,
was as good as could be defired. Equal attention
has been paid to other branches of provifion. Our
officers are not a little vigilant in taking care that
all articles are in due prefervation, and of the proper quality.

Some alterations might, however, be ftill made,
with advantage. The allowance of falt meat, is
too much at fea: when a fhip leaves the harbour,
it ought to be reduced one third, and the full

* As foap has not yet been fully introduced to fupply the
people, the captains would do well to infert an article into their
private orders, for the men to buy it at pay-day, and to be muf-
tered with their cloathing at ftated times.

value

value of it fupplied in fomething elfe. I think fome of the cheaper pickles would be very acceptable. The molaffes by admiralty order being now made general, the oatmeal breakfaft is rendered palatable, and muft be highly relifhed; it forms a valuable article in the vegetable part of the diet. In the Weft Indies, cocoa is fupplied for breakfaft with fugar, in lieu of fome fpecies of provifion liable to fpoil in that country. The late Captain James Ferguffon, lieutenat-governor of Greenwich Hofpital, was the benevolent inventor of it.

Our fhips are now liberally fupplied with frefh vegetables for the firft fortnight after coming from fea; by which means, along with frefh meat, they are quickly recruited for any emergency, and prepared to refift the effects of falted beef or pork for a longer time. I think if we could give every fhip, on going to fea, a quantity of onions, to mix with their pea-foup, or even to feafon the falt meat, it would be ufeful in preventing fcurvy. It is much to be wifhed that a few fheep fhould be carried out, for the ufe of the fick: a little mutton broth is fo nourifhing under debility, and fo defirable in many cafes after a long cruize, that to grant it would be the *ne plus ultra* of our improvements. Our officers have kindly fhared their ftock with the fick; but look at their pay, alas! they cannot afford it.

E 2   Two

From experiments which I made at Weevil, in the fummer of 1792, when furgeon of the Duke, a certain means was found to preferve *water* pure and fweet for any length of time; by the fimple procefs of firing the cafks, in putting the ftaves together, till a *charry coat is formed over the whole furface.* They are to be found in a little work intitled, Medical and Chemical Effays; printed by Jordan, in Fleet-ftreet, London. I am not, however, fatisfied with the attention the Victualling Board has given to this method; and it has never yet been practifed from fufficient authority.— Other obfervations and propofals for improving the diet, will occur in different parts of the work, among the difeafes.

CONCLUSION.

It has for fome time been a cuftom in the navy, for Captains to give out a code of regulations, for the obfervance of officers and men in their fhips. Many of thefe are of great importance in the prefervation of health, and nearly connected with our fubject. Officers, however, have different ideas of modes of difcipline; and we fee the condition of fhips, and the conduct of the people, differ materially. It would appear invidious in me, to felect or point out fuch as I might think moft worthy of preference and imitation.

tation.   But I cannot help being of opinion, that
a fyftem of valuable regulations, for the internal
government of his Majefty's fhips, might be
compiled from them.   I am acquainted with fome,
where, in cafes of fire in any part of the fhip,
the diftribution of the people, and what ought to
be done, are detailed with the moft perfect experience of a practical art ; and at the fame time
with the moft correct knowledge of thofe branches
of philofophy, which are required on a fubject of
that nature.   Others I know to be equally exact
in all that relates to cleanlinefs, to wafhing decks,.
to the treatment of the fick, and the fituation of
a fick-birth, which is moft commodious under the
forecaftle, to take in the round-houfe, and to be
perfectly free from the gally fmoke.   A Captain,
fingularly attentive to his fick people, afked me
one day, if it was not improper to wafh white
paint, though in a fick birth, with vinegar: he
always obferved a dampnefs afterwards, which was
occafioned by the vinegar converting the cerufe
into a fugar of lead.   This was certainly good
chemiftry ; and white paint in fick births fhould
be always wafhed with foap and hot water.—In
the London, Vice-Admiral Colpoys has ordered
the books fent to the fhip by the focieties, to be
kept in the fick birth, for the amufement and
inftruction of his people : others have checkers,
&c. to keep the feamen employed: and I re-

E 3   member,

member, Captain Charles Thompfon ordered the
people of the Vengeance *tops*, by way of exercife,
when we approached the cold weather, on our
paffage home.

———————

During the prefent war, many obfervations have
occurred to me, which point out fome alterations
to be neceffary in the payment of fmart-money to
the feamen. When the cheft of Chatham was
firft inftituted, that was confidered the chief naval
port : but from the immenfe increafe of our navy,
fince that time, and the bufinefs to be done at
Portfmouth and Plymouth, it becomes exceedingly inconvenient for the men to attend there ;
and the intention of the charity is often fruftrated.
Now, there would be other good effects reaped by
the fervice, were this money paid at the other
ports. It is given with the benevolent view of
alleviating the diftreffes of men, who have been
maimed on duty ; and it would be of material
confequence for the officers of the fhip to attend,
that the character of the man might be better
certified, than can be done in a fmart-ticket ; for
furely there muft be occafion, now and then, for
difcrimination ; as, when the wound was received
on any hazardous undertaking. It would thus
operate as an encouragement, and would be held
up as a reward of merit and great exertions,
whether

whether in an engagement, or other species of duty. I also think that an admiral and two captains, would be the fittest officers to award these favours of Government to the deserving objects. Their station must afford them the best opportunities to appreciate the value of the service; and by carrying their surgeons with them, or the physicians of fleets and hospitals, and surgeons of hospitals, the qualities of the hurts would be better ascertained. But a sailor would receive the boon with double gratitude, when it came from the hands of officers high in rank; and it would convince them, that they had an interest in doing him justice, and in his future support and happiness.

I would have it preserved as a sacred *Creed* in our Navy, never to separate the condition of the seaman from his officer: who can feel for men of this description, like the Commander, to whom he he must look for protection, whether in health or disease? It is this that begets the reciprocal attachment among them, and hence the numerous instances of a paternal solicitude for their welfare, which are daily coming under my review.

MEDICINA NAUTICA.

GENERAL ABSTRACT

OF THE

STATE OF HEALTH IN THE FLEET,

FROM THE 1st OF JAN. 1794, to DEC. 1796.

AND OF THE WEATHER DURING THAT PERIOD.

I THINK it neceſſary to begin my remarks from this time, as my attendance at Haſlar Hoſpital afforded me ſome opportunities of knowing the general ſtate of health in the Fleet.

It is not intended in the following pages to give a minute detail of the changes of weather, as have been uſual in ſome medical regiſters; but only to mention thoſe particular occurrences in each ſeaſon, which evidently tended to affect the health of the fleet. Journals of the weather, which are given at length, ſwell the work unneceſſarily; and I believe are not always read. It is, therefore, my intention to be very brief on this part of my ſubject.

The

The winter of 1793-4, was rather mild upon the whole; not remarkable for the quantity of either rain or snow that fell; and there were but few days of frost.

The fleet had been cruizing late in the season; and by chasing a French squadron of seven sail of the line, a considerable distance to the westward, were by an easterly wind, which lasted upwards of three weeks, prevented from getting into the Channel, before the 20th of December.

The Ruffel and Invincible, at this time, or very soon after, landed a number of men ill of a contagious fever. So late as the 19th of March following, two very bad cases of this fever were sent to Haslar Hospital, from the Ruffel. The sick from this ship complained much of washing decks so very often during cold weather; it was performed no less than thrice a week; and the poor fellows attributed their sickness to this cause: certainly, with great reason, it ought to be considered as having materially assisted the effects of infection.

In February, some cases of small pox were sent on shore from the Alfred. The infection was traced to a child brought on board in its mother's arms. They, in general, did well.

I may here mention a contagious fever which appeared on board the *Raisonable*, commanded by Lord Cranstoun.* This ship had been long at

---

* This ship did not belong to the fleet.

sea,

fea, and had met with much bad weather. She
came to Spithead in January, and landed upwards
of a hundred, very ill of typhus and dyfentery.
Mr. Newberry, the furgeon, fufpected the contagion to have been brought on board by draughts
of men from the guard-fhip, but did not trace it
to any particular perfon.   During their abfence
from England, one hundred and feventy people
were fent on fhore at one of the weftern iflands,
under Mr. Newberry's care.   Lord Cranftoun
humanely attended to every propofal from a judicious furgeon, for making the fick comfortable,
and they were fupplied with every thing neceffary.
Of this number, ten died.

A confiderable number of cafes, in flux and
fever, were received from the Gibraltar.   This
fhip had failed from Plymouth in November, to
join Lord Howe; but after a cruize of fome weeks,
during much ftormy and rainy weather, not falling
in with the fleet, fhe was obliged to return to
port.   Of thefe a few died with fymptoms of great
malignity.

Thefe fhips being all affected with fevers very
much alike in their nature, in their progrefs and
fymptoms, probably received the contagion by
men fent to them from receiving fhips and tenders,
which during fine weather attracted little notice,
but appeared under a more ferious form, when the
cold and wet weather fet in.   None of them had
been

been long in commiffion; but from any inquiries which I made, nothing fatisfactory could be learnt.

Feb. 23. This day fix men, in typhus, were received at Haflar Hofpital, from the London. On examining the particulars of their fituation, I was not a little furprized to find, that fome of them had been bled two or three times; and a larger quantity of blood taken, than is ufual among feamen, even under inflammatory complaints. They were, as might be fuppofed, in a ftate of extreme debility. I was not able to trace this fever to its true fource, but had reafon to believe that it much refembled the condition of other fhips, juft mentioned. Thefe people, and others who followed them, to a man complained inceffantly of the fevere duty they had undergone in returning the fhips ftores to the dock-yard. They were not difcharged on the eighth of April, when I left Haflar; to fuch a degree of weaknefs were they reduced.

It having appeared to me, from the treatment of the fever, as now narrated, that the officers of the London were not aware, that a difeafe of a contagious nature was extending itfelf among their crew; I therefore requefted Dr. Johnftone, the refident commiffioner of fick and hurt, to make the inquiry. Dr. Johnftone was told, that no difeafe of this defcription was known on board the London. I

have

have been thus far minute in my remarks; as the
fequel will fhow, that there were juft reafons for
my early apprehenfions.—In the mean time, the
London was paid off, and her complement diftributed to other fhips.

March. The Valiant of feventy-four guns, Captain Pringle, having received four hundred men
from the London, fome of them are now come on
fhore, in typhus.

About this time, an infectious fever prevailed
on board the Hebe frigate, Captain Alexander
Hood. She fent fome cafes to Haflar, with fymptoms of uncommon malignity. This fhip was
always remarkably leaky in her upper works; and
the moifture between decks, in bad weather, did
not fail to affect the people. Lieutenant Minto of
the marines received this fever, from vifiting
fome of the privates in the wards at Haflar.   Mr.
Leggat the furgeon, was timely aware of the infection; but it was not fubdued without fpreading.   The frigate had been long in commiffion,
which is a circumftance always favourable to our
means of preventing fever; but the feafon of the
year was unfavourable *.

March

* " In March 1794, the real febrile infection was introduced into the fhip by fome men we received from a fhip
from the coaft of Guinea. A number of our men died of it,

at

March 10. This day ten men in fever, came from the Valiant. Some have also come from the Cæsar and Leviathan, part of the London's crew.— Among these come from the Valiant, is a boy that had been some time on board. He attended the surgeon's mates in their mess, and was frequently sent on errands, by his masters, to the sick birth, where he received the infection.

March 16. In the course of the week fifty people have been sent from the Valiant.

March 18. Fives cases of typhus received from the Cæsar.

April 1. Cases of fever continue to be sent from the Valiant, Cæsar, and Leviathan.

About this time, the Africa, of sixty-four guns, Captain Home, arrived from Plymouth, and sent some bad cases of fever to Haslar *. I took the liberty, from being well known to Captain Home, of mentioning the circumstance, and he had no suspicion that a fever of this kind was on board: but it was easily traced to eleven men, who had been discharged from the Cambridge receiving-ship,

---

at the hospital. It was some time before we got rid of the infection, although we kept fires constantly between decks, and smoaked the ship frequently with tobacco, wet with vinegar, brimstone, &c.

        (Signed)   JOHN LEGGAT, Surgeon,"
Hebe, Jan. 1795."

====

* The Africa did not belong to the Channel Fleet.

fhip, at Plymouth, before the Africa failed. Captain Home gave orders, inftantly, for every man
with the flighteft fymptoms of fever, to be feparated and fent on fhore. The people who were
not fufficiently flopped, received additional cloathing; and the ftricteft attention to cleaning both the
perfons and cloathing of every man, was immediately put into execution : all the bedding was
fpread out to air, and fires put in every place,
wherever moifture or foul air could be generated.

One of the flag-officers then at Spithead, having heard my opinion of the Africa's fituation,
and putting no great confidence in it, was pleafed
to order two furgeons to report to him their ideas
of the condition of that fhip. Thefe gentlemen
infpected the fhip, and alfo her people at the
hofpital; but were of opinion, that the fever was of
little confequence: they alfo told the admiral,
that the treatment of thofe at Haflar, was very
improper, and could not fail to kill more than it
would cure. It is to be remarked, that the firft
cafes, which were by far the worft, were now convalefcent under my prefcriptions. I fufpect, that
what they meant by an improper method of cure,
was my allowing to a few, three pints of port wine,
in the twenty-four hours. This opinion gave me
no uneafinefs; it plainly evinced, that thefe gentlemen were equally confined in their information
and experience on the fubject of typhus.

Captain

Captain Home, who had more confidence in my ſtatement of the caſe, did not mind any additional trouble on his part, when the health of his ſhip's-company was at ſtake. He perſevered in his means of clearing and preventing contagion; and in the ſpace of ſix weeks, after ſending near a hundred men to the hoſpital, had the ſatisfaction to ſee his labours crowned with ſucceſs.   Mr. Cudlip the ſurgeon, from his active humanity in the duties of his ſtation, during this ſtate of the Africa, became a ſufferer, and obliged to quit his ſhip when ſhe ſailed, with the ſquadron under Rear Admiral Murray, for Halifax.

On the beginning of April, or towards the end of March, a few bad caſes of fever came from the Robuſt and Coloſſus; both of which had arrived lately from the Mediterranean, in good health. No particulars came to me, on the manner of infection, but it did not extend-far.   Captain Jenkins was then commanding officer of the Coloſſus, and paid ſingular attention to my directions : there was no ſurgeon preſent.   Theſe people having come from a warm climate, complained, with juſtice, of the hulk they lived in; which was left in a very wet and dirty condition by the ſhip's company that had occupied it before.

It is worth while to remark, that during my attendance at Haſlar, and paying ſome attention to the progreſs of infection in the tainted ſhips,

4   I obſerved

I obferved that it conftantly varied, in clear and rainy weather. After a few days of wind and rain, the cafes of fever were always increafed in proportion. The fact would feem to be accounted for in this manner : during hard gales or heavy rain, the people, when confined below, muft breathe a more impure atmofphere ; which alfo may concentrate the contagious effluvia: and if employed and expofed to the cold and wet weather on deck, thefe will act as debilitating powers, and favour the action of contagion on the body. This obfervation was not new to me, I had feen it before on more occafions than one.

———————

April 9. This day I received the appointment of Phyfician to his Majefty's Fleet, under the command of Admiral Richard Earl Howe; dated at the Admiralty-Office, April 3, 1794.

The Fleet, at this time, confifted of thirty-two fail of the line, eight frigates, a floop, a firefhip, one cutter, two luggers, with the Charon Hofpital fhip, in which I was ordered to embark. On the fixteenth, I had vifited the fhips, and reported their ftate, as annexed, to the Commander in Chief *.

* State

* State of HEALTH on board his Majefty's Ships under the Command of ADMIRAL EARL HOWE, at Spithead, April 16th 1794.

SHIPS.	Number on the Sick Lift.	Objefts for an Hofpital.	Prevailing Difeafes.	Confined to bed.
Queen Charlotte	14			
Royal Sovereign	31	—	Slight Fever.	
Royal George	25	—	Venl Complts.	
Barfleur	20	—	Ditto.	
Impregnable	30	4	—	3
Queen	18		—	3
Glory	40	1	—	2
Bellerophon	21	1		2
Hector	19	—	Catarrhs.	1
Montague	50		Venereals.	
Tremendous	25	2 Surgical Cafes.		2
Gibraltar	13			
Valiant	20	6	Typhus.	6
Ramillies	17			
Audacious	12			1
Brunfwick	26	—	Venereals.	2
Cæfar	32	5	—	5
Alfred	at St. Helens.			
Defence	32			
Ganges	40	—	Venereals.	2
Leviathan	45	—	Ditto.	1
Majeftic	12	—	Catarrh.	5
Bellona	11	2	—	1
Invincible	16			
Arrogant	46	—	{Mealles and Catarrh.	14
Orion	22			
Ruffel	7			
Marlborough	5			
Thefeus	5			
Alexander	15			
Thunderer	33			
Culloden	28	—	—	3
	at St. St. Helens.			
	725	21	—	53

FRIGATES.

Phæton
Latona
Hebe
Venus
Niger
Pallas
Southampton
Pegafus
Charon, H. S.
Circe
Aquilon
Comet, F. S.
Oreftes brig
1 Cutter, and
2 Luggers

} No Sick on Board.

Thofe marked * left the Fleet before the glorious 1ft of June.

T. TROTTER, M. D.
PHYSICIAN.

F

The weather at this time, was mild, as is ufual in the month of April. The Valiant was ftill infefted with the fever that was brought on board by the crew of the London ; but the cafes were becoming milder ; in fome, it put on an intermittent form, which fhowed, that it was now on the decline. In the Leviathan, it was now compleatly extinguifhed. This fhip received the marines of the London. In the Cæfar, fo late as the fifteenth, I found five or fix of the London's people, confined to bed with fever; they were fent immediately to the hofpital ; and at this period, it might be faid to ftop *.

The

* "There is a diminution of the fever this month, nor are the fymptoms fo malignant : it is ftill, in a great meafure, confined to the people we received from the London ; there being but few exceptions to the contrary ; I believe not more than three or four of our old fhips company have been attacked with it ; and one or two of the Raifonable's. A draught of forty or fifty men, from that fhip, were put on board of us the morning we left Spithead, to compleat our complement. *

Out of ten of our people, returned from Haflar, cured of the fever, four have relapfed ; three of whom were again fent to the hofpital, with every fymptom greatly aggravated. Several have been cured on board, as we have not always had opportunities of fending them out of the fhip. I have generally treated them in the following manner : I gave an emetic

of

* Mr. Seeds does not feem to have known, that a typhus fever had but lately prevailed in the Raifonable.

The Arrogant, with a new-raised crew, had just come from the Nore. The decks and beams of this ship were remarkable for their moisture; which seemed to exhale from the timber. The purser, an intelligent man, had been ill from rheumatism, which he ascribed entirely to this cause; and had for some time observed the damp vapour, on his bed-cloaths, every morning like a heavy dew. It had also affected the ship's company; and a kind of irregular intermittent was prevalent among

of ipecac. when they first complained; after which, they took diaphoretics, combined with pectorals; and they were occasionally ordered a scruple or half a dram of one of the neutral salts, with a few grains of rhubarb, as throughout the disease they complained of being constipate d.—I found they could not bear evacuations, except of the most gentle kind; even emetic tartar, in the smallest dose as an emetic, ruffled them exceedingly. I continued the diaphoretics and pectorals, till the tightness across the breast, pleuretic stitches, and difficulty of breathing, were in a great measure removed; at this time the pulse became softer, and an expectoration was produced. Blisters, I found serviceable in relieving the rheumatic pains and stitches. The cure was then compleated by tonics and opium. I have given in this stage of the complaint, brandy and water, as a substitute for wine: wine not being at present allowed to the sick *.

Cammomile tea, joined with the bark, I am inclined to think, answers better than bark alone, in this fever."

Cæsar. Mr. SEED's *Report for April*, 1796.

---

* The Commander in Chief, about this time, had ordered the surgeons to demand wine from the pursers, when wanted.

T. T.

among the landmen on board.   This uncommon
dampnefs of the Arrogant's timbers, was probably very much owing to the fpot where fhe lay
in ordinary, which was furrounded with fwamps.
But, whether it was owing to this caufe, or the
fhip being compofed of green wood, it was an
unpardonable omiffion, fomewhere, to report her
fit for commiffion before fhe had been fired, for a
length of time fufficient to deftroy the moifture.
The meafles alfo prevailed in the Arrogant, at
this vifit ; but were mild, and foon difappeared.

When confulted by the Admiral, on the propriety of taking the Valiant to fea, I gave it as
my opinion, that in the prefent declining ftate of
the contagion, fhe was fit for fervice.   I added,
that the fine weather, which we had a right to
expect from the feafon of the year, and the employment and activity which the duty of the fhip
would give the people, would all tend to accelerate the extinction of the remaining contagion.
I knew, befides, that much was to be trufted to
the method in which Captain Pringle was accuftomed to difcipline his people, and the unwearied attention and ability of Mr. M'Callum,
the furgeon.   The Valiant, therefore, failed with
us.   Some relapfes of men that had been fent
from the hofpital, happened at this time.   It was
imputed to the fpare diet which they had lived
upon in a ftate of convalefence.

May

May 1. The Valiant has had no fresh attacks for three days ; four are now confined to bed, and these but slight cases.—The whole of the fleet may now be said to be in perfect health.

2. The fleet got under weigh, with a strong breeze at N. E.

4. The convoy, with the men of war under Rear-Admiral Montague, were ordered to separate.

13. This day one man in small-pox, and another in measles, both in a state of eruption, were sent on board the Charon from the Gibraltar. These diseases did not extend to any other person.—Six seamen, and the gunner's child, were inoculated, in consequence, on board the hospital ship ; all of whom had the disease in the mildest degree.

17. The fever list of the Valiant is at present ten ; but so gentle are the complaints, that none are confined to bed. One man died since leaving port : he was a landman, and came on board only the evening before the fleet sailed *.

* " State of HEALTH, on board his Majesty's Ship Valiant, from the 1st of May to the 1st of June, 1794.

DISEASES.	Since last Report,				Present Sick List.
	Taken ill.	Sent to the Hospital	Recovered.	Dead.	
Fever - - -	47	—	41	2	4
Flux.					
Catarrhal Complaints	12	—	9	—	3
Rheumatism -	4	—	3	—	1
Scurvy.					
Venereal Complaints	18	—	13	—	5
Ulcers - - -	10	—	7	—	3
Bruises, &c. -	6	—	4	—	2
Total - - -	97	—	77	2	18

REMARKS.

The fever that has for some time past prevailed among our people, has of late been diminished, and seems now nearly worn out. Those cases of it which have occurred latterly, have been slight, with a very general intermittent tendency. Ever since its first appearance, few other complaints, comparatively, have occurred; and these few have been of the catarrhal kind; in some degree combined with the fever; in general, however, easily cured, but equally infectious with the fever in its more distinct form.——We have had no appearance of scurvy, although some of our men had lately before we sailed, come off a long cruize, during which they had been afflicted with it. This, I hope, is owing to the very judicious supply of lemon juice and sugar.

June 1st, 1794.
(Signed) GEORGE M'CALLUM, Surgeon."

19. This

19. This day Earl Howe received intelligence that the French fleet was at fea; when the fignal was made for the fhips to prepare for action.

Our courfe was therefore fhaped for the ene-my's fleet, which was cruizing, to protect their homeward-bound American convoy, a long way to the weftward. During this purfuit, a number of veffels belonging to our Lifbon and New-foundland trade were retaken, with two cour-vettes and a cutter; all of whom were immediate-ly fet on fire, and the people taken out.

27. Two French prifoners, in fever, were fent to the Charon from the Thunderer. One of them died next day.—The infection did not extend to any of the Thunderer's people : they came from a courvette, which was bearing difpatches to the French admiral*.

28. The enemy's fleet were feen to windward.

JUNE 1. The detail of this glorious battle, fo honourable to the Britifh navy, will fall to the fhare of fome more able hiftorian. We were left in poffeffion of feven fail of the enemy's line, one of whom, the Vengeur, funk the fame evening.

FRENCH SHIPS TAKEN.

		killed.	wounded.
Le Sans Pareil -	80 —	260 —	120
Le Jufte - - - -	80 —	100 —	145
L' America - - -	74 —	134 —	110
L' Achille - - -	74 —	36 —	30
Le Northumberland ..	74 —	60 —	100
L' Impeteux - - -	74 —	100 —	75
		690	580
Le Vengeur - - -		320 funk in her.	

* Dr. Blair, in his letter of this date, mentions his opinion

RETURN of the Killed and Wounded on board his Majefty's Ships, in the Actions with the French Fleet, on the 29th and 30th of May, and of the 1ft of June, 1794.

SHIPS.	KILLED.		WOUNDED.		TOTAL.
	Seamen.	Marines.	Seamen.	Marines.	
Cæfar - -	18	0	37	0	55
Bellerophon -	3	1	26	1	31
Leviathan - -	10	0	32	1	43
Sovereign - -	11	3	39	5	58
Marlborough -	24	5	76	14	119
Defence - -	14	4	29	10	57
Impregnable -	7	0	24	0	31
Tremendous -	2	1	6	2	11
Barfleur - -	8	1	22	3	34
Culloden - -	2	0	5	0	7
Invincible - -	9	5	21	10	45
Gibraltar - -	1	1	12	0	14
Queen Charlotte	13	1	24	5	43
Brunfwick - -	32	12	94	20	158
Valiant - -	1	1	5	4	11
Queen - -	30	6	57	10	103
Orion - -	5	0	20	4	29
Ramillies - -	2	0	7	0	9
Alfred - -	0	0	6	2	8
Ruffel - -	7	1	24	2	34
Royal George	18	2	63	9	92
Montagu - -	4	0	13	0	17
Majeftic - -	3	0	4	1	8
Glory - -	13	0	31	8	52
Thunderer -	0	0	0	0	0
Audacious - -	4	0	18	0	22
Phæton frigate	3	0	4	0	7
Total	224	34	699	109	1,098

Names

of thefe cafes; and his reafons for a fpeedy removal to the hofpital, to fecure their own people from the contagion.

Names of OFFICERS Killed and Wounded.

KILLED.

Ships.	Officers.	Qualities.
Sovereign -	Mr. W. Ivey - -	Midſhipman.
ırough - -	Abm. Nelſon -	Do.
:e - - -	Wm. Webſter	Maſter.
	John Fitzpatrick	Boatſwain.
;nable - -	David Caird -	Maſter.
ndous - -	Francis Roſs -	Firſt Lieut.
Charlotte -	R. Rawlence -	Seventh do.
	John Neville -	Lieut. Queen's Reg.
- - - -	W. Mitchell -	Maſter.
;eorge - -	G. Heigham -	Eight Lieut.
	John Hughes -	Midſhipman.
ıu - - -	James Montagu, Eſq	Captain.
- - - -	Mr. Geo. Metcalfe	Maſter.
	David Grieg -	Midſhipman.
ck - - -	Alex. Saunders	Capt. 29th Regt.
	Thomas Dalton	Maſter's Mate.
	James Lucas -	Midſhipman.

WOUNDED, and unable to come to Quarters.

ıon - -	T. Paſley, Eſq. - -	R. Adm. of the White.
	— Smith, Eſq. - -	Capt. Marines.
	Mr. Chapman - -	Boatſwain.
n - - -	Glen - - -	Midſhipman.
ıvereign -	T. Greaves, Eſq. -	Adm. of the Blue.
	Mr. C. Money - -	Capt. Marines.
	S. Mitchel - -	Lieut. do.
ıugh - -	Hon. G. Berkeley -	Captain.
	Mr. A. Ruddock -	Second Lieut.
	M. Seymour -	Fifth Lieut.
	Fitzgerald - -	Midſhipman.

WOUNDED, and unable to come to Quarters, *continued.*

Ships.	Officers.	Qualities.
Marlborough *continued*	Mr. Shoreland - -	Do.
	Linthorne - -	Do.
	Hon. M. Clarges * -	Do.
	Mr. M. Pardoe - -	Mafter's Mate.
Defence - - -	Mr. J. Elliot - -	Do.
	Boycolt - -	Enfign Queen's Reg.
Impregnable - -	W. Buller * - -	Lieutenant.
	Patterlo - -	Boatfwain.
Barfleur - - - -	G. Bowyer, Efq.	R. Adm. of the White.
	Mr. W. Prowfe - -	Sixth Lieut.
	Fogo - - -	Midfhipman.
	Clemons - -	Do.
Queen Charlotte † -	J. Holland - -	Do.
Queen - - - -	J. Hutt, Efq. * -	Captain.
	Mr. Dawes * - -	Second Lieut.
	Lawrie - -	Sixth do.
	G. Aimes - -	Acting do.
	Kinneer - -	Midfhipman.
Ruffel - - - -	Stewart - -	Do.
	Kelly - - -	Do.
	Douglas - -	Boatfwain.
Royal George - -	J. Ireland * - -	Second Lieut.
	J. Balmborough	Mafter.
	Boys - - -	Midfhipman.
	Pearce - - -	Do.
Montagu - - -	Hon. Mr. Bennet -	Do.
	Mr. J. Moore - -	Do.
Brunfwick - - -	J. Harvey, Efq. * -	Captain.
	Mr. R. Bevan - -	Lieut.
	Hurdis - -	Midfhipman.
	Harc. Vernon -	Enfign 29th Reg.
Culloden - - -	Trift. Whitter -	Third Lieut.
Invincible - - -	W. Whithurft -	Midfhipman.

Thofe marked * died of their wounds afterwards.

† Captain Sir A. S. Douglas was wounded on the forehead, but returned to the deck after being dreffed by the furgeon: a fmall exfoliation of the frontal bone was the confequence.

SURGEONS

SURGEONS of the Fleet on the 1st of June.

Queen Charlotte - - - -	Mr. Wm. Murray.
Royal Sovereign - - - -	Dr. Alexander Young.
Royal George - - - - -	Mr. Richard Shepherd.
Barfleur - - - - -	John Heath.
Impregnable - - - - -	Wm. Wallis.
Queen - - - - - -	Alexander Browne.
Bellerophon - - - - -	Thomas Fargher.
Tremendous - - - - -	Alexander Dods.
Gibraltar - - - - - -	Dr. George Smyth.
Valiant - - - - - -	Mr. George M'Callum.
Ramillies - - - - -	John Plumpton.
Brunswick - - - - -	Robert Forrest.
Audacious - - - - -	Robert Mellville.
Alfred - - - - - -	John Birtwhistle.
Defence - - - - - -	Dr. James Malcolm.
Leviathan - - - - -	Mr. Charles Boveard.
Cæsar - - - - - -	Thomas Seeds.
Invincible - - - - -	Thomas Kenning.
Orion - - - - - -	William Pattison.
Ruffel - - - - - -	Joseph Stephenson.
Montagu - - - - -	Samuel M'Clure.
Majestic - - - - -	Leonard Gillespie.
Marlborough - - - -	Thomas Romney.
Glory - - - - - -	Peter Smith.
Culloden - - - - -	Robert Ramsay.
Thunderer - - - - -	Dr. Primrose Blair.
Phæton - - - - -	Mr. Andrew Baird.
Latona - - - - - -	James Turkington.
Niger - - - - - -	Robert Kirkwood.
Southampton - - - -	George Michie.
Venus - - - - - -	John Buchan.
Aquilon - - - - -	Robert Harris.
Pegasus - - - - -	William Fuller.
Charon, H. S. - - - -	William J. Warner.
Comet, F. S. - - - -	Robert Caruthers.
Incendiary, F. S. - - - -	
King's Fisher, S. - - - -	
Rattler and Ranger cutters -	

June 8th. The Admiral having been informed
that the French prizes were fickly, ordered me to
vifit them as foon as the fituation of the Fleet
would permit.

The number of killed and wounded in thefe
fhips was very great. The account, as given
above, was what I obtained from the French furgeons, and reported it to the commander in chief.
They were alfo fickly; and, from being dirty to
an extreme degree, a contagious fever had carried
off many. The Sans Pareil had particularly fuffered. This fhip failed from Breft, in a fquadron,
fix weeks ago, with a complement of a thoufand
men on board: many perifhed in the early part
of the cruize, and a hundred ill of the fever were
fent into a courvette, whofe guns were taken out,
for the better accommodation of the fick to be
carried to port. It was told me, that the lower
deck ports had never been opened from leaving the harbour till the day of action. The deaths
in confequence were very great; but we could not
come at the exact number. Lieutenant Jacob,
of the Majeftic, had the command at this time,
whom I cautioned about the nature of the fever,
that his people might be kept as much as poffible
from the infected. This worthy man fell a victim
to his humanity, with two midfhipmen.* They

* What a pity, that men endued with fuch noble feelings
did not live to fhare that reward, by promotion, which this
nice difcharge of an officer's duty would have fo juftly confer-
red upon them!

superintended some of the French prisoners in cleaning the lower deck, where the sick and wounded lay; and there they received the fever, of which they died at Haslar Hospital.   Captain Trowbridge, who had been taken by this squadron, in the Castor, with part of his company, were on board the Sans Pareil.   I examined some of the Castor's men, who were infected: Captain Trowbridge was administring to them himself, with great concern for their fate, and attended me to their hammocks.

The Northumberland was the, next in point of sickness; but the others were more or less infected with the typhus fever.*   The whole were well supplied with medicines, surgeons, and assistants. The necessaries carried to sea for the use of the sick, consisted of many articles of comfort unknown in our navy: they had even live stock † on board for this purpose; which, when bestowed on our sick, must come from the tables of their officers.

June 9th. Nothing remarkable in the state of the weather happened during this memorable cruize.

A prisoner in the small-pox was this day sent to the hospital ship from the Valiant, as soon as discovered.   No infection followed in the Valiant.

13th. This day we arrived at Spithead with our prizes; part of the Fleet, under Admirals Greaves

---

* The Vengeur, which sunk soon after the action, was also infested with a contagious fever.

† Sheep, goats, and poultry.

and

and Caldwell, having gone into Plymouth. The wounded men were landed immediately.

It deferves, to be recorded to the immortal honour of the officers in this Fleet, that the flock of their meffes, confifting of fheep and poultry, with all the delicacies which their tables afforded, was cheerfully refigned for the fupport and comfort of the wounded. But their goodnefs did not flop here; they learned that the diet of the hofpital was deficient in fome articles which a wounded failor could wifh for; and fums of money were fent to procure them. My heart warms with indefcribable emotions, while I relate a fact that deferves to be recorded with the pen of an angel! *

Towards the end of June the weather became extremely hot, and the thermometer rofe to 80 of Farenheit in the fhade.

A few days after the arrival of the Fleet at Spithead, Mr. Mackie, Surgeon of the Southampton, informed me that a fever of a contagious nature

---

* Wihle the officers were extending their bounty to the wounded feamen, reciprocal feelings, like an electrical fhock, pervaded every corner of the country. A fubfcription was begun at Lloyd's coffeehoufe, by fome public fpirited Merchants, which was quickly imitated by all the great towns in England and Scotland, and foon filled, to the amount of *betwveen twenty and thirty thoufand pounds!* The Theatre Royal, Drury Lane, where the company performed *gratis*, is faid to have contributed 1,800l.

The wounded officers and feamen, marines and foldiers, received large donations: and a vaft fum was *funded* for the fupport of widows and orphans, under the benevolent direction of Mr. Devaynes and others.——Such are the bleffed effects of a naval triumph!

nature feemed to fpread among his people; and he fufpected that it was communicated by their intercourfe with the French prifoners. This gentleman being perfectly aware of the danger, had conducted himfelf with fingular addrefs and ability: the infected were carefully feparated from others as foon as the flighteft fymptoms of fever appeared, and on the arrival in port they were fent to the hofpital. By thefe timely precautions not more than fifty were infected; the latter cafes were trifling compared with the firft, of whom one or two died. Thus the Southampton may be faid to be the firft fhip cleared while at Spithead.

Being fully fenfible of the danger to which the greateft part of our fhips had been expofed, from a general contagion, I did not wait for reports coming to me from the furgeons on the progrefs of the infection; but made it my bufinefs to fee the whole. In the Majeftic and Ramillies, particularly, fome bad cafes appeared, with every fymptom of malignity; alfo in the Barfleur, Cæfar, Valiant, Bellerophon, Alfred, and Invincible. I have formerly mentioned, that the Majeftic had taken poffeffion of the Sans Pareil, and that the communication between them rendered her crew more liable to the fever.

In the beginning of July the weather continued ftill fultry to an uncommon degree.

Another frigate, the Circe, had been very early infected at fea, by means of fome of the prifoners

4

taken

taken out of the courvette, before the action on
the 1st of June; but in this ship it made but
little progress: Mr. Dunn's report only mentions
twelve; and these are not said to have been attended with much danger.   From the nature of
the fever being quickly detected, the Circe's people were soon secured against its future attacks.*

The infected ships, during the month of July,
were employed in the means of stopping the contagion.   Those officers who had confidence in
fumigation, performed it every morning: the
decks were kept clean, and the whole inside
white-washed: constant attention was paid to the
cleanliness of the people's cloaths, and the bedding
was spread abroad every day to air.   The seamen
were ordered to keep themselves clean in their
                                                   persons,

* "During the early part of the cruize there were few complaints of a serious nature.   Several prisoners from a French
courvette, and other prizes, were sent on board the Circe; two
of which were soon attacked with symptoms indicating an
alarming fever.   Some of our best men became daily affected
in a similar manner; want of appetite, nausea, a foul tongue,
some degree of tension of the skin, with costiveness, were the
most observable appearances. Emetics were administered early
in these cases, and blood-letting avoided where there were no
violent symptoms of inflammation.   Purgatives were necessary
to clear the intestines, and relaxation of the surface was kept
up by antimonials, sometimes combined with opiates.   Three
or four days illness was sure to induce great debility.   A free
use of wine was highly relished by them all; and they soon
gained their usual strength.

                    (Signed)

Circe, June 1794.   ROBERT DUNN, Surgeon."

perfons, and to fhift more frequently.   Fires were
kindled in pots in the hold, well, and bread-room;
ftoves in the orlop, cabletiers, and fore and after
cockpits.   Care was taken that the circulation of
air through the wings fhould not be interrupted;
and, befides the common windfails, two ftunfails
were fitted for the fore and main hatchway, fo
that every corner below was pure and compleatly
perflated by the air; fome of thefe fails were kept
trimmed during the night, fo as to counteract the
effects of the heat when the ports were down.
A quantity of vinegar in an iron pot was frequently, in the courfe of the day, converted into
vapor, by plunging a red hot loggerhead into it, or
fprinkled over the decks.   Such were the method
which our officers perfifted in; befides, the cables,
fails, cordage, with all kinds of ftores, were brought
upon deck during the day, and the ftore-rooms, in
the mean time, were ventilated and white-wafhed,
fo that not a particle of impure air could lodge any
where.   The immediate feparation of the infected
was, however, what I moft depended upon; and not
even the flighteft cafes were allowed to remain in
the fhip.   The furgeons were directed to order
their mates to walk frequently round the deck;
and to watch thofe men who, while fitting in
their berths, appeared dejected or folitary.   This
was always my own practice in an infected fhip;
becaufe feamen will fometimes withhold their

G

complaints

complaints for a day or two, under the idea that an incipient fever is only a common cold *.

Some officers, not without reafon, were extremely alarmed at this general infection in the Fleet ; and they apprehended accumulated danger from the very fultry weather. The ftate of the weather was, however, with me, a caufe of confolation : and a few who watched my opinion on the fubject, with fome fufpicion, did not fail to confefs, in a very fhort time, that their apprehenfions were removed.

The Valiant had been for fome weeks clear of her former infection, when this fever made its appearance. It fell, however, into proper hands. Mr. M'Callum detected the firft fymptoms of it among the marines, who had been centinels over the prifoners under the half deck : other furgeons, with equal accuracy, difcovered it firft among the quarter-mafters, or thofe who flept near the edge of the hatchways that led to the hold where the French were confined.

The Queen Charlotte was faved from this fever, by fending all her prifoners with the leaft fymptom

of

---

* I have mentioned *white-wafhing*, among our preventive means : it is, I believe. a general practice, and fuppofed to be ufeful by attracting *carbonic acid gas*, from the atmofphere. In our *hygeine*, we; however, employ it as a part of cleanlinefs, and as giving to the decks a cheerful pleafant appearance : it will, in this manner, infpire agreeable fenfations from light and delicacy, and may have its fhare in the fcale of general ftimulants, to fortify the body againft debilitating powers.

of indifpofition, to the number of feventeen, on board the Charon, at fea.   One of thefe patients died next morning, with every fymptom of malignity.   How far every fhip might have availed themfelves of this falutary precaution, I cannot pretend to fay; but it ought to be attended to in future; and an hofpital fhip provided, fufficiently large to accommodate the fick of a large Fleet, on fimilar emergencies.

The Royal George did not feem to fuffer much from the infection.   Mr. Shepherd, in his report for Auguft, mentions thirty-three fevers, only ten of whom were fent to the hofpital; they affumed a bilious remittent form, probably partaking of that type from the feafon.*

In the Ruffel alfo, in the month of Auguft Report, Mr. Stephenfon mentions a flight fever, that had not been known during the preceding month.   Now, this fhip and the Royal George, were

* " During this month we have had more fevers than at any former period; moftly of the bilious remittent kind; which, in general, terminated favourably, by clearing the ftomach and bowels of their crude and acrid contents, which produced a remiffion; when the bark, wine, and a reftorative regimen, compleated the cure.

The cafes (10) fent to the hofpitals were inclined to degenerate into the low fhip fever.

(Signed)

Royal George, Aug. 1794.   RICHARD SHEPHERD, Surgeon."

G 2

were probably infected, by one of our own ships, during the frequent communication, while in harbour. *

While we were employed in clearing the Channel Fleet of this fever, a vast mortality is said to have prevailed among the French prisoners, in the neighbourhood of Portsmouth and Plymouth. Not only the nurses and attendants on the prisoners, at Forton and Hilsea, but even the Middlesex Militia, who did the duty of guard over them, were affected; some of whom died. I have not been able to learn, what, means were practised to check the ravages of this fatal distemper, among those men whom the fortune of war put into our hands. There was employment enough for me at Spithead; but I have been told, that three hundred and thirty-eight Frenchmen died at Forton alone, in less than four months.

From the arrival of the fleet, on the 11th of June, to the final extirpation of this fever, 800 cases were sent to Haslar Hospital, of whom forty died, which is one in twenty. This mortality was, in every ship's crew, confined to the earliest stage of infection: out of one hundred and forty-six, sent by the Barfleur, seven died the first fortnight: the last hundred sent, were slight, and soon recovered. A few were sent from the ships of

* " The fever appeared of the slightest kind; but the ship being fitting for sea, they were sent to the hospital as soon as the symptoms appeared.

(Signed)

Ruffel, Aug. 31, 1794.   J. STEPHENSON, Surgeon.'

of the fleet at Plymouth, to that hofpital, but I have not obtained the return. We fufpect that a winter contagion would have produced a very different degree of mortality from what we fee here; and alfo a more permanent difeafe.

Towards the end of July, our endeavours had been fo far fuccefsful, that I did not hefitate to pronounce the tainted fhips fit for fea. What cafes appeared at this time, were either with diftinct remiffions, or the fymptoms fo moderate, as to yield to trifling remedies. In the Glory it was protracted from particular caufes*. Sir G. K. Elphinftone hoifted his flag on board the Barfleur, and a draught of a hundred men was fent from one fhip to the other. The brave crews of both fhips thus parted with their officers, and with one another. Never did poet or painter delineate the pangs of feparation in a more affecting manner, than was exhibited in this exchange of fhips. There was a fault in not moving the whole. Of the hundred men to be fent from the Barfleur, to replace thofe of the Glory, were fome ill of the fever, or convalefcent at Haflar. The depreffion of fpirits, on being told that they were to gc to another fhip, occafioned immediate relapfes in fome, while others, for a time, feemed to fink under it. A degree of contagion was, however,

* The Glory was fitted for the flag of Sir G. K. Elphinftone before the cruize, but this officer did not join the fleet at that time.

carried

carried to the Glory, by this means: but never
did the attachment of a Britiſh ſailor to his ſhip-
mate and officers, appear under more heroic
tokens of friendſhip!   It was truly worthy of the
1ſt of June! *

IT

---

* EXTRACTS OF REMARKS FROM SURGEON's
REPORTS.

" There was nothing particular appeared in the fever of this
month, until the twenty-third; when we received ſome men
belonging to the Alfred, who had been ſome time on board the
Juſte French prize, with a few belonging to our ſhip.   Three
of the Alfred's men were taken ill the day after they came on
board, and two of ours.   It was evening when they complained: they each had an emetic, and next morning were
ſent to the hoſpital   The ſymptoms were chiefly great debility, with pain in the back, and redneſs of the eyes, with a
wildneſs in their looks, and tremor of the tongue, &c.

(Signed)

Invincible, June 1794.   T. KENNING, Surgeon."

" Five caſes of typhus have occurred during this month;
in my opinion, they were cauſed from contagion, in conſequence of having on board a number of French priſoners.
They were ſent to the hoſpital.

(Signed)

Glory, June 1794.   P. SMITH, Surgeon."

N. B. During the greater part of the time that the fever prevailed in the Barfleur, Mr. Heath, the Surgeon, was on ſhore,
attending Admiral Boyer, who had loſt a leg in the action;
conſequently no reports were ſent to me.   This ſhip ſent more
men to the hoſpital than any other. *T. T.*

" The

IT having appeared to me, during the late operations of the fleet at fea, that the diet of the hofpital fhip was extremely deficient, in articles
of

" The Alfred had been remarkably healthy, previous to the French prifoners being fent on board; the fever was imported by them. Every means have been ufed to ftop its progrefs; in which we have now, in a great meafure, fucceeded.

(Signed)

Alfred, July 1794. JOHN BIRTWHISTLE, Surgeon."

." The fever has, for the laft three or four days, appeared to abate, in the frequency of cafes; and the fymptoms have become more mild. (Signed)

Ramillies, July 20th, 1794. JOHN PLUMPTON, Surgeon."

" Towards the latter end of this month, the attack of the fever became lefs violent, and generally feized thofe men that had been on fhore for fome days : * but from a fufpicion that the contagion was not entirely gone, I fent them to the hofpital fhortly after they complained.

(Signed)

Invincible, July 1794. T. KENNING, Surgeon."

* Mr. Kenning, with great difcernment, alludes here to the excefles, which feamen are apt to commit when on fhore. When there is infection in a fhip, officers ought carefully to prevent intoxication, and to keep their men conftantly under their eye. For this reafon, no perfon fhould be fuffered to go on fhore; befides the hazard of carrying the difeafe to other fhips.—Vide Mr. M'Callum's Remarks, under the chap. Typhus. *T. T.*

G 4

" The

of comfort for the fick, which induced me to
apply for fome addition; the Lords of the Admiralty were pleafed to comply with my application.   We were now enabled to carry with us
                                                ftock

" The fever which has for fome time paft been predomi-
nant, appears at this time to have nearly fubfided.   The
fymptoms in thofe at prefent attacked, are not near fo violent
as formerly, when it appeared under a very malignant form.
The debility was alfo great on former attacks : the eyes were
much flufhed, the tongue black and crufty, with thirft, and in
fome cafes with immediate delirium : the pulfe in general was
quick and feeble.   The fever was alfo remarkable, by being
attended with fore throat, * and confiderable pain in the fhin
bones.   No fuch fymptoms at prefent appear.   Every attention is paid to cleaning and airing the fhip.
                       (Signed)
Bellerophon, June 1794.   J. B. HOUSEAL, Surgeon."

---

     * Quere! Might not this unufual complaint of the throat,
be the globus hyftericus; a fymptom which I have myfelf,
very frequently obferved in the incipient ftage of typhus ?
                                                T. T.

" The Majeftic's fhip's company were tolerably healthy from
the time of the Fleet leaving port, the 27th of April, until
ten days after the action of the 1ft of June, as the Sick
Lift never exceeded in number thirty-five, during that
period ; and as there were feldom more than two or three
fick confined to their hammocks at that time.   A catarrhal
fever, whofe type was remittent, and character benign, was
the epidemic.   It feldom continued longer than three days,
going off by fweat, or expectoration ; in fome the parotids
                                                were

ftock and vegetables, fruit, pickles, eggs, porter, &c. and to purchafe milk, when in port.

Auguft 16. This day I vifited a fquadron of Portuguefe fhips, confifting of five fail of the line, a frigate,

---

were confiderably fwelled; in others, the fauces were inflamed. In the fcorbutic and pulmonic, it occafioned a troublefome cough and expectoration, which continued for fome weeks, and which required blifters, expectorants, and in fome, bark and opiates.

On the memorable 1ft of June, we had not a man from his quarters, on account of ficknefs; but on that and the following days, having received upwards of two hundred French prifoners from the Sans Pareil, moft of whom were highly fcorbutic, and many of them had been attacked by an infectious fever, epidemic, on board that fhip, for feveral weeks: a very great alteration foon took place in the health of our fhip's company.

The caufes of ficknefs among the people of the Sans Pareil, as fuppofed by the furgeon and officers, were attributed,

1ft. To the fhip's company being moftly compofed of land-men, who neglected to keep themfelves clean:

2dly, The vapor of bildge-water; the cocks for watering the well in French fhips, being now fuppreffed:

3dly, Humidity of the fhip: being newly conftructed, and no means being ufed to obviate the ill confequences of humidity below, by means of fires, wind-fails, &c.

4thly, To the intemperature of the weather, which, during a two months cruize to the weftward, they had experienced to be unfettled, dark, foggy, and humid.

The

a frigate, and two brigs, under the command of
Rear-admiral De Valle, who were come to join the
channel fleet. It did not appear to me, on
walking round the different decks of these ships,
that

The French prisoners confined in the hold, in calm, sultry,
foggy weather, were soon attacked with fever, in considerable
numbers. The origin of it, the subjects it attacked, the extreme
debility, despondency, universal lassitude and pains, dejection of
countenance, foetor of the breath, irregularity of type, and the
eruption of petechiæ, or marbled appearance on the skin, all
shewed the malignant nature of the disease: which, with
Sauvages, may be termed, *febris nautica* : with Huxham and
others, *febris putrida, maligna, petechialis* : with Burferius de
Kanefeld, (Institutiones Medicinæ Practicæ) who, in oppo-
position to most of our English authors, and I think with
justice, looks on it as an exanthematous fever, *sui generis*;
morbus petechialis : and to which opinion the people accede,
terming it the *purple fever.* *

The infected prisoners were removed, as soon as possible, on
board the Sans Pareil, where they died in great numbers.
About the 10th (of June) the disease began to make its appearance among the Majestic's ship's company : the approach
of the *solstice*, and the *canicula*, seemed to have considerable
influence on the spreading of the complaint; as, notwith-
standing

---

* These spots are not cutaneous eruptions, they are not to
be felt on sliding your finger over the surface; they are not a
disease of the skin, but small effusions of venous blood, in the
cellular texture, which are seen through the integuments :
when a number of them run together like stripes, we call
them *vibices*. Dr. Darwin says they are owing, as in scurvy,
to diminished venous absorbtion. *Zoonomia.*   T. T.

that they had many fick on board; but in all of
them I could perceive people loitering about in a
ftate that did not indicate perfect health. None
of them were confined to bed, yet there were
evidently

---

ftanding the precautions ufed, many were attacked with it.
Having fent fixty-eight perfons to the hofpital; at prefent its
violence feems diminifhed, as we have only fent three within
the laft four days.

With regard to the treatment of this difeafe, in fuch patients
as were kept on board; bleeding was not indicated in any I
have feen. A gentle emetic of ipecacuanha, or infufion of
chamomile flowers, given in the beginning, had good effects,
and followed by a fudorific opiate, as theriaca, at night, and
wine and bark the fucceeding days, appeared to check the
progrefs of the complaint in its origin.

Keeping the belly open, by injections and tammarind decoction, feemed to anfwer very well. Neutral falts, as nitre
joined with camphor, feemed to have falutary effects. Blifters,
in the advanced ftage of the complaint, when the head and
breaft were much affected, had evidently the beft effects.—
When the degree of fever was not high, and diftinct remiffions
took place; when the determination to the head and breaft
was not great, the tongue moift, and the belly not conftipated,
the bark, though it did not always arreft the fever in its courfe,
was attended with good effects, by fupporting the ftrength of
the patient, and obviating a tendency to putrefcency: the
fame may be faid of wine, prudently adminiftered. Opium
and antimony, except when ufed as a diaphoretic in the beginning, as is mentioned above, appeared to me to be attended
with difadvantageous effects: the firft occafioning, or increafing delirium, coma, extreme debility, conftipation, &c.;
and

evidently among their complaints, the milder degree of fhip-fever.   When I confidered the nature
of the climate to which thefe people were accuftomed, and the change which they muft foon experience,

---

and the latter debilitating the patient very much, bringing on
diarrhœa, fuppreffion of expectoration, conftant naufea, &c.

The eruption of petechiæ, or marbled appearance on the
fkin, preceded by pain of the back and loins, oppreffion at the
præcordia, fighings, pain and heavinefs of the limbs, and irregular weak quick pulfe, generally fhowed itfelf from the 4th
to the 7th day.——The difeafe generally went off about the
14th or 17th day, moft commonly by perfpiration; whilft the
appearance of *concoction* took place in all the excretions.

A copious expectoration proved *critical* to fome of thofe
who had been affected with pulmonic complaints: a diarrhœa
appeared to be *critical* with others; but, in general, it was
fymptomatic, and attended with danger.

(Signed)

Majeftic, July 20, 1794.   LEONARD GILLESPIE, Surgeon.[12]

" It will appear, by comparing the prefent month's ftatement of health with the preceding, that the infection, which
was at that time prevalent, appears now to be totally obliterated; only eight being fent to the hofpital this month, and
thefe but little indifpofed.

(Signed)

Bellerophon, July 1794.   J. B. HOUSEAL, Surgeon."

" On my joining the Glory, I was informed by my prede-
ceffor, that fhe had been always very healthy; and, particularly,

rience, towards the fall of the year, in our channel fervice, I could not help telling Sir Roger Curtis, Captain of the fleet, that the firft gale of wind, and wet weather, in the failor's phrafe, would *knock* the whole of them up.

Auguft

---

larly, that fhe had efcaped a fever, with which many of the fhips had been infected, after returning from the late cruize. However, as fhe had received a number of men from a fhip much afflicted with fever, there was every reafon to expect, that fhe might foon become tainted with a fimilar difeafe. This, in fact, happened; and it has continued ever fince. Although none have died on board, or at the hofpital, as far as I have learned, it has neverthelefs been very diftreffing, and continues to rage, notwithftanding the greateft attention to get it under. When it had the appearance of getting better, I imagine it gained frefh vigour, either from a man returned from the hofpital with a high degree of fever upon him; or, from the apparel of the men brought out of the filthy bed-houfe at Haflar, which, I believe, has been often the fource of contagion.*

(Signed)

Glory, Auguft 1794.   GEORGE M'CALLUM, Surgeon."

---

* My benevolent friend had been five months in infected fhips; and hence his caufe of complaint. A governor, with three lieutenants, being now in that hofpital, on a liberal allowance from government, it may be expected, that the cloaths and bedding will not only be frequently taken out of the houfe and aired, but that every fpecies of apparel will be returned to the fhips, with the men, all pure and well wafhed, as they ought to be.   *T. T.*

Auguft 22. This day the whole of the fleet dropped down to St. Helen's; confifting of thirty-four fail of the line, befides frigates.

While the fleet lay at St. Helen's, there were fome cafes of fever fent from the Royal Sovereign and Orion, to the hofpital-fhip. Thefe fhips had refitted at Plymouth, after the action of the 1ft of June, and had been but a fhort time at Spithead. A feaman belonging to the Orion died. The whole of our patients were fent on fhore, in order to keep the hofpital clear for fea-fervice.

September 3. The fleet failed with the wind at N. N. E. September 5th, fell in with the home-ward-bound Eaft India convoy, on a different tack, in the night. From the darknefs of the night, fome of our fhips run foul of the Indiamen, one of whom was difmafted: this obliged the Ad-miral to bear up for Torbay.

September 7. The fleet failed from Torbay.

From the 18th of September to the 22d, we experienced fevere gales of wind from the weft-ward. Many of our fhips fuffered; but the Portuguefe fquadron was in extreme diftrefs, fome of the fhips having loft foremaft and bow-fprit. This induced the Admiral to return to Torbay.

September 23. This day I vifited the fhips of the Portuguefe fquadron, by Earl Howe's order,

and

and reported their situation to him in the evening. The purport of my report was nearly as follows: " They are all sickly, and a contagious fever is prevailing among them. Since the bad weather at sea, they have rapidly increased in the number of their sick."—There is so great a contrast between the appearance of one of these ships, and an English man of war, that to go from the one to the other, was like coming from a sepulchre to a banquet. Not only the men, but even the officers themselves, are accustomed to lie down in bed with their clothes on : the beds of the people are spread along the deck, and not slung in a hammock, as with us. At the time of my visit, much trouble seemed to have been taken to give their sick an appearance of delicacy, for they were all laid on new beds, with clean shirts on. But I well knew, that under such circumstances, and nasty habits, as laying upon deck, neither health or cleanliness could be preserved. Contagion must therefore spread among them, with uncommon rapidity. In a future part of this work, I shall mark some peculiarities of constitution, in the inhabitants of warm countries, that seem to favour the action of typhous infection, when they come to a northern latitude. On this account I hinted to the Admiral, that it would expedite the recovery of the squadron, to return immediately to their own country : but this was impracticable

at

at the time, from their difabled ftate. The Gama,
the flag-fhip, had one hundred and twenty fevers
in her lift; none had yet died, but there were many
dangeroufly ill.

In the Reine de Portugal, commanded by the
Marquis de Niza, an accomplifhed young noble-
man, five had died. The lift was increafing faft.
I had opportunities of feeing fome, with the firft
rigours upon them, attended with great apprehenfions for their fafety, and an extreme dejection of fpirits. The other fhips were more or
lefs in the fame fituation. This fquadron was
ordered to Plymouth to refit, and purify themfelves from a general infection.

It is worth while to remark, that this fever exactly correfponded with one which infefted a
fquadron of her faithful Majefty's fhips, which
came to Spithead the preceding year; many of
whom were fent to Haflar Hofpital. *

September

Charon, September 23, 1796.

* " S I R,
   In my report to Admiral Earl Howe, con-
cerning the ficknefs prevailing in the fquadron of her faithful
Majefty's fhips under your command, in Torbay, I have had
occafion to ftate, that the fever is of a contagious nature, and
likely to gain ground as the cold weather approaches. Many
cafes are, no doubt, flightly affected; but there are others with
every fymptom of malignity and danger.

" I have alfo ftated, that according to your cuftom of fervice, the fick and found are laid on deck; a practice highly
improper,

September 27. The hofpital-fhip having received fo much damage during the late heavy
gales of wind, and her ports not being water-tight,
it was deemed expedient to fend her to Plymouth
to refit. The whole fick of the fleet, amounting
to feventy, were fent on board the Charon to be
carried to the hofpital. Thefe confifted of a few
cafes of phthifis, one of fmall pox, fome of chronic
rheumatifm, ulcered legs, and eight invalids, from
the Barfleur.

At this time there was not a fingle fever in
the fleet, that could be faid to have originated
from contagion, or the leaft remnant of that
fpread from the French fhips. Thus ended an

H   infection,

---

improper, as it expofes them to the moifture of the deck, muft
create filth, and generate and extend contagion.

" I have ftrongly recommended a more liberal ufe of wine;
but I know that your furgeons have very different ideas, from
their theories of fever.

" I am at a lofs to recommend any methods of prevention
under the circumftances of your fhips; and have only hinted
to the Commander in Chief, that the health of your people
very much depends on the fpeedy return to a climate more
adapted to their conftitutions than the cold of our winter.

I have the honour to be,

S I R,

Your moft obedient,

and very humble fervant,

To

T. TROTTER.

REAR ADMIRAL DE VALLE, &c.

infection, the moft general that was ever fpread in a large armament; and which, under other fituations of feafon and difcipline, might have proved fatal to the channel fleet of Great Britain. The fpeedy extinction of it will reflect ever-lafting credit on the officers of thofe fhips where it appeared; and their exertions will remain a fine model for their fucceffors to imitate, under fimilar circumftances of fervice.

I have remarked, that a cafe of fmall pox was fent on board the Charon; but the fhip's name, from whence he came, has been omitted in my Journal. It was in the eruptive ftage, and the difeafe affected no others where he belonged. Having given orders to inquire whether our patients had all paffed through the fmall pox, it was found that three never had the difeafe. In a former part, I mentioned that fix people and a child had been inoculated in the Charon, during the fummer cruize. Thefe patients were now ordered to be inoculated as foon as matter could be procured from the puftules of the other man. This was, however, neglected by the Surgeon's mate: while at Plymouth Hofpital, the fmall pox appeared in the natural way, of the confluent kind; and I am forry to add, two out of the three died.

October. The Portuguefe fquadron having arrived in Hamoaze, to refit and purify, opportunities

tunities were now given me to learn ftill farther
the iffue of the contagion which appeared among
them in Torbay.

The Europe, a third rate in ordinary, was
allotted to them for an hofpital, and fitted accordingly; but from the numbers daily taken ill,
in all the fhips of the fquadron, fhe became fo
crowded, in the fpace of three weeks, that no
room was left to receive more. It was during
this diftreffing condition, that Admiral de Valle
requefted me to meet the Phyfician to Plymouth
Hofpital, on board the Europe, to examine the
ftate of accommodation, as well as that of the
people, and to report to the Lords Commiffioners
of Admiralty, the rapid extenfion of this prevailing fever, among the feamen of the fquadron.

After waiting two hours beyond the appointed
time for the phyfician of the hofpital, I propofed
Dr. Mein, of the Caton hofpital fhip, as an
affiftant on this neceffary, though dangerous duty.
To this the Doctor very cheerfully confented; it
would, indeed, have been cruel to refufe attendance, as the fafety and comfort of fo many human
beings depended much on our furvey and ftatement of the cafe.

There were, at this time, on board the Europe,
five hundred people, in different ftages of the fever.
The lower gun-deck had, at firft, been the only
part appropriated for the fick; but it could not

contain

contain two hundred cradles without being crowded; other parts of the ship were therefore occupied. We found the fick fufficiently clean in their perfons, and beds alfo : but, as I have formerly remarked, there was fome doubt that this was always the cafe ; and it could not be expected, that in fo fmall a fpace, either diet or attendance could be complete. The orlop deck being full in every corner, from its very imperfect ventilation, was literally peftiferous. The fmell was intolerable ; we walked round the fore and after cockpits; but were not able to make many inquiries, or to attend to every particular; indeed, refpiration could not be immediately accommodated to an atmofphere, not only deprived of much of its oxygene, but ftrongly charged with contagious matter. The moaning and ghaftly looks of the whole; the skin fallow, livid, and black; *petechiæ & vibices*; tremors, *fubfultus tendinum*, and in a few general convulfions ; groans of the dying, and thofe in pain; raving of the delirious ; tongue black, parched, and tremulous; the breath fœtid; refpiration quick and laborious ; the eyes funk, fixed, glaffy, moiftened with tears; were among the prominent features of this hideous groupe of human mifery ! The convalefcents were kept on the upper deck, but from being much expofed to cold and wet weather, many of them fuffered relapfes.

6 From

From this furvey, the crowded ftate of the Europe was reprefented to the Admiralty Board : another fhip was immediately ordered by their lordfhips to accommodate the fick of this unfor‧ tunate fquadron, and one of their own fhips was appropriated for convalefcents alone.   What pro‧ portion of deaths happened I have not been able to learn : but the mortality here, as in all fimilar fituations which I have witneffed, was greateft among thofe feized at the early period of infec- tion, when the difeafe appeared under the moft malignant form.   Thefe fhips had fo far recover- ed about the middle of February, that they failed for Lifbon ; leaving behind, on board the Europe, near three hundred men, under the care of Dr. Mein. *

The

* REPORT TO THE ADMIRALTY.

" S I R,

            The commander in chief of the Portugueze fquadron, in Hamoaze, having requefted me to furvey his people on board the Europe ; yefterday, accompanied by Dr. Mein, of the Caton hofpital fhip, I performed that duty.

" Admiral de Valle, having requefted us to report the con- dition of his fick, for the purpofe of laying it before the Am- baffador of his Court, in London; I am alfo defired to make known to you, for the information of the Lords Commiffioners of Admiralty, the ftate of thefe people.

" In addition to my report, made to Admiral Earl Howe, when ordered to infpect thefe fhips at Torbay, I have only

H 3   now

The precautions which I chiefly recommended to Admiral de Valle and his officers, were the immediate separation of the infected; plenty of warm cloathing; no wafhing of decks, but in lieu of it, dry rubbing, or fcraping; wind-fails for the orlop deck; fires to be kept conftantly burning, for preferving their fouthern conftitutions warm; *bunting* fafhes for the lower deck ports; the utmoft attention to cleanlinefs, both of perfon and cloaths;

now to obferve, that what I then fufpected, has been too early verified; the fever having become progreffively worfe, From the number daily taken ill, the Europe is fo crowded, in every corner, that much danger muft arife from fo many people, in a difeafe highly contagious, being confined in fo fmall a fpace. Means cannot therefore be too foon devifed for thinning their prefent quarters, and preparing other apartments for their accommodation.

" The number of deaths is now confiderable; and cafes in the moft dangerous ftage of the fever, bear a very large proportion. We have recommended every method for prevention, ufual in fuch fituations: but fo much in the prefent inftance depends upon peculiarity of conftitutions, difcipline, and habits of the people, that we cannot flatter their Lordfhips with the profpect of a fpeedy extinction of the contagion.

I have the honour to be,

SIR,

Your moft obedient,

And very humble fervant,

T. TROTTER.

To
PHILIP STEPHENS, Efq.
Admiralty Office.

cloaths; a pint of wine *per diem*; no expofure to moifture; and laftly, the utmoft exertions in officers to keep the minds of the people engaged, fo as to infpire confidence and cheerfulnefs, and to divert them from thinking on the danger of infection. Thefe forms, I believe, were as ftrictly attended to, as the nature of the difcipline would admit. The worthy officer who commanded the fquadron, was moft fenfibly afflicted with the fufferings of his people, and was borne down with the calamity when I firft met him at Plymouth: but I had the fatisfaction of feeing him recover his fpirits. From this time the general health of his fhips became better.

The Charon having undergone what repairs were found neceffary, returned to Torbay, on the 3d of November; bringing with us what recovered men were fit to join their fhips.*

We

* The day after I vifited the Europe, being employed in preparing the report for the information of the Portugueze Ambaffador, I was more or lefs affected with head-ach. As I was about to fit down to dinner, at three o'clock, a meffage came from Rear-Admiral Bourmafter, on board the Glory, in Caufand Bay, defiring to fee me, as he was very ill. I went on board the Glory immediately; and, from being much fatigued at my return, went early to bed. About four in the morning I waked, extremely agitated, with ficknefs at ftomach, fhivering, &c. I now lived with Dr. Mein, on board the Caton, and defired the fervant to call him. He agreed

H 4   with

We now received the fick of the Fleet, chiefly the difeafes of the feafon, catarrhs, rheumatifms, &c. and with them on board we proceeded to Portfmouth. The Fleet returned to Spithead about the 24th of November.

The weather, during the months of October and November, was mild as to temperature, but thick and hazy, with much rain, and hard gales of wind from the fouth-weft quarter; which obliged the Fleet to put into Torbay repeatedly for fhelter.

There had been a total exemption from fcurvy fince we left Spithead; owing, in a great meafure, to a general fupply of fugar and lemon juice, diftributed to the fhips on leaving port in the end of Auguft.

Early in December a fevere froft fet in, which continued, with frequent falls of fnow, all the month.

---

with Mr. Worgaen, that an emetic would be proper, left the attack fhould be from typhus contagion. I took fome ipeca-cuanha wine, which excited fufficient vomiting; and felt no more of the complaint. Fatigue, and expofure to cold, were probably the caufe of my indifpofition.

— 1795. —

The froft continued equally fevere all the month of January, and the thermometer fell frequently fo low as 22° of Farenheit : fome days it blowed extremely hard, with fnow and hail, from the N. E.   About this time it was reported that the French fleet had put to fea in great force, with the view of intercepting our outward-bound Weft India convoy.   This report was foon confirmed by Sir Sidney Smith, in the Diamond frigate, going into Breft, and hailing a fhip of the enemy's line, that had returned difmafted from their fleet to the weftward.

Jan. 26th. The Fleet having affembled, Earl Howe failed from Spithead, with thirty-three fail of the line, and ten frigates, &c.

The wind coming to the weftward, when off Plymouth, the Fleet put into Torbay, on the 1ft of February.

The Cumberland, of feventy-four guns, arrived at Spithead on the beginning of January, from the Nore, and newly fitted.   Her fick lift, which was one hundred and nine, furprized the Admiral, that he doubted of her ability to go to fea in fuch a feafon.   On my firft vifit, the lift had increafed to one hundred and twenty ; but I found

none of them confined to bed; and their complaints were all of the catarrhal kind. What
tended, perhaps, in a great degree to aggravate the
difeafe among this fhip's company, was the receiving a great part of her complement from the
Vengeance, juft returned from the Weft Indies;
and confequently the conftitutions of her crew were
more fufceptible of cold. A few of the worft
cafes were fent on fhore; but I did not find the
fymptoms, in general, run fo high, as to make me
hefitate a moment whether the Cumberland was
in a condition for actual fervice. Captain Rowley was very active in procuring them what additional cloathing the weather indicated, and his
people, for the cruize, did very well.

This catarrh had all the appearances which
ufually characterize it when epidemic. While
the Fleet remained at Spithead, the only fhip befides the Cumberland affected with difeafe, was
the Coloffus. The number of her fick, on the
weekly account, induced the Admiral to order me
to report her fituation. It was a prevailing catarrh. Some of her men were confined to bed,
with a confiderable degree of fever, and ftricture
about the breaft: the worft of the whole were
fent on fhore to the hofpital.

While the fleet remained in Torbay, the weather was very unfettled; fometimes a fouthweft
wind,

wind, with rain, at other times northweft, with
fnow; on the whole it was bleak, wet, and cold.

Before the 5th of February, the Catarrh was
more or lefs general in every fhip of the fleet.
This epidemic, on former occafions, has been traced to have gradually fpread itfelf from the
fhores of the Baltic, or the Low Countries, to all
the fouthern and weftern provinces of Europe,
The firft fhip affected in our fleet, the Cumberland, came from the eaftward, the Nore, where
alfo other fhips at the fame time complained.
It is not improbable but a certain conftitution of
the atmofphere is neceffary to produce it, and
give it vigour, although it is afterwards fpread from
one perfon to another, Its general progrefs in
the fleet, in fo fhort a time, juftifies this idea;
for there were fhips with the difeafe on board,
where no communication with any other could
be fufpected.—In the year 1782, a fimilar catarrh
prevailed in the Channel Fleet, then in the North
Sea, under the command of Earl Howe.   It
fpread with inconceivable rapidity over the whole.
The fleet was obliged to return to port, and it
was fome time before the difeafe difappeared: a
confiderable number of deaths was the confequence.   A fquadron of eight fail of the line
lay at that time in Torbay, under the command
of Rear-admiral Kempenfeld.   The influenza firft
made its appearance on board the Fortitude of
feventy-

seventy-four guns. Mr. M'Nair, the surgeon, informed me, that only a few men complained in the evening: but so quick was its progress throughout the ship's-company, that next morning when all hands were called to weigh the anchor, more than two hundred said they were sick, and could not come to the capstan. The Fortitude was therefore ordered to Plymouth, to send her people on shore: but the other ships that went to sea, were obliged to return, nearly as bad of the same disease, two days after.

When the catarrh was epidemic in 1782, I was surgeon of the Buftler sloop of war, at Plymouth: I was affected with the disease in a severe degree. In the present case I was also a sufferer, and had a relapse, from exposing myself too soon in the boat, to rain and cold weather, in Torbay.

February 14. The fleet sailed from Torbay. In two days the immense convoy for the West Indies, Lisbon, Mediteranean, &c. were able to get out of Hamoaze, when the Admiral made sail to the westward.—In the mean time the French fleet had returned to port in great distress; with the loss of five ships of the line, and other damages, from gales of wind.

February 20. Off Cape Finisterre.—The weather uncommonly mild and fine. Therm. 52°.

February 25. Arrived at Spithead.

The

The fleet was now almoſt free of the influenza; only one hundred and ſixty men, in various diſ- eaſes, after a month's abſence, were ſent on ſhore at Haſlar.   The number of deaths during this epedemic, was twenty-eight.

A fever, of the typhus kind, appeared among the gentlemen of the Invincible in Torbay, which extended to a number of caſes.   Yet it is ſingu- lar, that it did not affect any of the ſeamen, al- though ſome of them attended their officers, who ſlept in the gun-room, and where communication with the ſhips-company was not prevented *.

March

" In the laſt week of January, two gentlemen in the cock- pit were taken ill with a bad fever. The reaſon that I did not take notice of it in my report for that month, was, that it did not appear to me to be of much conſequence ; but on the 2d. inſt. when the third was taken ill, I had not a doubt but it was contagious.

Every precaution was taken to keep the gentlemen ſeparate from one another : the worſt were put into the gun-room. The attack was inſidious for ſome days.   They complained of chills, nauſea and head-ach : one had a ſevere pain over the left eye, which laſted for a few hours, and then ſhifted to the other, and ſo on alternately.   They took emetics on com- plaining, and the bowels were kept regular ; afterwards the pulv. antimon. and about the fifth or ſixth day of their illneſs, they took bark and wine, with a full doſe of opium at bed- time.   In one caſe the opiate did not procure reſt ; he paſſed a better night with only wine.   I did not find that a more early uſe of wine was in any caſe detrimental ; it was given

night

March 1ſt. The following STATEMENT, taken from the monthly reports of the ſurgeons, will give a general idea what the fleet experienced from the catarrh in the ſpace of ſix weeks ; from the 18th of January to the 28th of February. It ought to be remembered, that what are mentioned here, are only thoſe, who from the degree of indiſpoſition, were admitted into the ſick liſt ; flighter caſes are not taken notice of.

Influenza.		Influenza.		FRIGATES.
Orion - -	480	Hector - -	150	Melampus - 53
Hannibal - -	314	Culloden - -	90	St.Margarita 21
Bellerophon -	104	Brunſwick -	180	
Cæſar - -	130	Royal Sovereign	160	La Nymphe 70
London - -	272	Marlborough -	80	Hebe - - 58
Barfleur - -	450	Leviathan - -	80	
Queen Charlotte -	No	Venerable -	270	Aſtrea - 30
Report from her Surgeon.		Invincible -	84	Thalia - 40
Impregnable -	200	Robuſt - -	20	
Zealous - -	180	Cumberland -	260	Aquilon - 64
Valiant - -	234	Queen - -	12	Creſcent - 44
Audacious -	130	Royal George -	30	
Ruſſel - 122 in port		Excellent - -	20	
Tremendous -	60	Canada - -	160	
Coloſſus - -	213	Triumph - -	80	
Prince of Wales	184	Gibraltar - -	20	
		Thunderer - No report from her Surgeon.		While

night and day with the bark. In one a bliſter was evidently of ſervice : the patient was inclined to delirium, when the bliſter was applied, and in the morning he was better, and recovered

While the fleet lay in Torbay, the Admiral was pleafed to order a number of fheep to be given to every fhip, in proportion to the fick lift, no frefh beef being allowed. This fupply came opportunely for our convalefcents, who fared fumptuoufly from it, and foon recovered ftrength. The mild weather during the paffage to Cape Finefterre, put an end to the epidemic.

March 6. In my general vifit to the fhips, after our arrival at Spithead, many recent facts were offered to my obfervation; which fupported the former arguments I have employed againft the charge of fifteen fhillings for the cure of the venereal difeafe. Thefe were of a nature too, that they could have been produced in evidence, had it been found neceffary. In fome of the fhips were men who undertook to cure the difeafe in

all

covered flowly afterwards. In the other cafe, blifters gave uncommon pain, and the patient was worfe all the next day. When he was fent to the hofpital, on the twenty-third day of his illnefs there was no appearance of this fever among the fhip's company.

(Signed)

Invincible, Feb. 28th. 1795.   THOS. KENNING, Surgeon."

Mr. Kenning mentions fifteen cafes of fever, in his report for this month; twelve of whom were cured on board, the other three were fent on fhore to Dartmouth fick quarters, and there recovered.   T. T.

all its ftages, and had more than one or two patients under their care. In others, the boats
crews were in the practice of bringing medicines
from the fhore to their fhipmates. A medicine
often fatal, even in the beft hands, was a popular
remedy with them. (Vid. " A Review of the Medical Eftablifhment in the Navy. Bew, London,
and Watts, Gofport, 1790.") This was no other
than Hydrar. Muriat. Some confulted itinerant
quacks, who flock to the fea-ports, and had paid
largely for their advice while fimple local complaints were converted into a confirmed lues.
Several had withheld the knowledge of the difeafe
from the furgeon, till the moft excruciating and
dangerous fymptoms had fupervened, and thus
became objects for an hofpital. On the whole,
there were abundant proofs at hand, for a purpofe, that I had long preffed with both head and
heart. I therefore refolved to addrefs the Commander in Chief, officially, on the bufinefs. My
reprefentation of the facts was received by Earl
Howe and Sir Roger Curtis, with all that warmth
of approbation, which the moft affectionate concern for the comfort of our feamen could dictate,
and for which the tranfactions of this fleet bear
ample teftimony. His lordfhip thought the fubject of fufficient importance to engage the attention of the Lords Commiffioners of Admiralty,
and laid it before them. The Board of Admiralty,
after making the neceffary inquiries as to the

amount

amount of the fum, in the furgeons pay, were pleafed to order an immediate ftop to be put to the
charge, and remunerated the furgeons by an allowance of money proportioned to the complement
in different rates. This alteration has been received, by the liberal and fcientific part of the
lift, with perfect fatisfaction. To thofe on foreign ftations, it is almoft clear gain, for few or no
venereal complaints prevail in the fhips on Eaft and
Weft India fervice. Thus terminated a perquifite,
illiberal from its inftitution, inhuman in its practice,
and impolitic from its continuance. It forms an
epoch in naval improvements ; for hundreds of feamen, have annually fallen victims to its effects *.

March

* EXTRACT of a Remark, from Mr. NICOLSON's Report
for March, 1795.

" Ever fince I have been a furgeon in his Majefty's navy,
I have obferved men labouring under venereal complaints very
backward in applying for cure ; often till they cannot walk
on deck in their watch. After tampering with themfelves for
a length of time, to evade a difgrace which they perfix to the
difeafe, they are only fit for an hofpital. They have alfo a
diflike to have fifteen fhillings appear againft their names in a
fhip's-book, and every finifter means is made ufe of to efcape
the ftigma. Hence a foundation is laid for a multiplicity of
abufes ; among others, complaints are feigned to make them
objects for an hofpital.

" Now Sir, if venereal charges were altogether abolifhed,
and fome recompence made the furgeon by way of medicine,
there can be no doubt but the men would come forward, to
get cured, at the firft appearance of the difeafe, and thereby

I

a great

March 15. The weather ftill continued cold,
and the winds for the moft part eafterly.   The
feverity of the winter had been general in this
country as well as on the continent : great numbers of cattle and fheep had perifhed in confequence, and vegetation was every where deftroyed.
The prices of provifions rofe in proportion, and
beef and mutton could fcarcely be procured. The
allowance of frefh meat to the fleet, was curtailed
to one ferving a week ; a difpofition to fcurvy became therefore general in our fhips.

I was early aware, from the concurrence of thefe
caufes, that a fcurvy muft foon appear in the
fleet, if not counteracted by other means.   For
thefe reafons, in order to keep the health of the
people ready for emergency, I propofed fome alterations in the diet.

1ft. In lieu of butter and cheefe, which of late
have been with difficulty procured, and oatmeal,

let

---

a great many ufeful feamen would be preferved, and the furgeon's duty unencumbered by a perquifite, not very pleafant
in its contribution as part of his pay.

(Signed)

Audacious.   E. Nicolson, Surgeon."

July 12th 1796. At the time I am writing this.; the
good derived from the abolition of the venereal fine, is uniformly teftified by every furgeon in the fleet. The men have
every where made their complaints fooner known, and there
are few fhips at Spithead whofe fick lift contains *fix* venereal
patients.   *T. T.*

let cocoa and fugar be allowed for breakfaft, as on Weft India ftations.   Or let fugar or molaffes be mixed with the bargou, or oatmeal gruel, that the ufe of them may become general, by being made palatable.   Query; why is oatmeal fupplied if feamen refufe to' eat it ?—Rice and fugar have been lately fubftituted for butter and cheefe, but the people diflike them.

2dly, Let the beer which is to be carried to fea, be made of double ftrength, with a larger quantity of hops to make it keep; and let two quarts be ferved inftead of a gallon of the fmall beer.   Which would make the expence equal, and not even occafion a fingle inconvenience in the forms of office.

3dly. Whenever the fhips return to port, let them be ferved abundantly with vegetables, in the frefh beef broth.   This is ftill a *defideratum* in the diet of feamen.   The quantity ought to be fpecified, and not left to the difcretion of a purfer, or any other officer, to withhold them altogether when they grow dear.   At this time there is not a cabbage brought to any fhip at Spithead.

4thly. Let fome of the cheaper pickles be fupplied; fuch as red cabbages, walnuts, cucumbers, &c. and let them be ferved in allowance with falt meat.   Thefe are cheap articles, and fine correctors of falted diet.   They are much to be preferred to four krout; grateful to the tafte, need no

wafhing

washing or cooking, and quench thirst when water grows scarce.

5thly. Let lemon juice be supplied to the surgeons for the cure of scurvy only. Cases of scurvy are now appearing in the fleet, and were they
supplied with lemon juice, the whole might be
cured on board.

Such was the first representation I judged it
necessary to make about the middle of March.
It was referred by their Lordships to the Commissioners of victualling, for their report; and I
was afterwards informed that the molasses were
ordered: but they did not come in time for
Lord Bridport's squadron to be supplied on the
11th of June; consequently the small squadrons
which sailed before that time reaped no advantage
from them.

March 17. Vice-Admiral Colpoys sailed with
the London, Valiant, Hannibal, Colossus, Robust,
and two frigates. Some of the lemon juice that
remained in the Charon, was given to these ships,
for I knew full well that they would need it.

About this time a squadron sailed for the Cape
of Good Hope, under Sir G. K. Elphinstone.
Some of his ships were taken from the Channel
Fleet, and had been exposed to the general causes
of scurvy, like others. On this account, as there
was no time to have supplies of fruit from London, I recommended Sir George to order lemons

to

to be bought on the fpot. This was immediately done, to the amount of fifty chefts, at the
fale of a cargo for the benefit of the underwriters *.

Some cafes of fcurvy foon appeared on board
the Minotaur and Invincible; and frefh ones were
daily added to other fick lifts, which made a
quick progrefs to the worft ftage of the difeafe †.
The honourable Captain Pakenham, with his ufual
generofity, fent fix guineas to his firft lieutenant,
to be expended on fruit and vegetables, as the
furgeon might think proper, as foon as he heard
that the fcurvy appeared among his people.

Finding the difeafe gaining ground in other
fhips ; I alfo confidered it my duty to reprefent
the ftate of the fleet to the Commiffioners of fick
and hurt ; and informed them, what appeared to
me to be the caufes of this general difpofition to
fcurvy throughout the whole of our fhips ; viz.
the fevere cold of the winter, the curtailed allow-

---

* To this precaution we are, in a great meafure indebted,
for the health of the fhips at the reduction of the Cape.

† " The fcurvy, within this laft week, has made its appearance, in a fevere degree, with thofe that have complained
of it : It is really a melancholly reflection, that feamen fhould
be attacked with fcurvy at Spithead, when the means that
would preferve them from that deftructive malady, are fo well
underftood. I fear that many of our fhip's company will
fhortly be laid up with it. As I have it not in my power to
cure them on board, I muft fend all thofe that appear to be
tainted to the hofpital."—Mr. KENNING's *Report for March.*

I 3   ance

ance of fresh meat, and the destruction of vegetation. At the same time I begged an immediate supply of lemon juice, or the fruit in its entire state, for the cure on board of those already complaining, and to prevent our ships companies from being *broke up*, by sending them to hospitals. Having given this information to the Commissioners, it was left to them to transmit this letter to the Admiralty, as they might think proper.

Most of the surgeons had a little lemon juice saved from the large supply in September last: this was given to the scorbutics while it lasted, and many were completely cured. When this was done, our men in scurvy were of course sent to the hospital.

April 16. A squadron under Rear-Admiral Waldegrave, consisting of the Minotaur, Invincible, Excellent, Tremendous, and La Nymphe and Blonde, were ordered to sea. The Excellent had many cases of scurvy, and a general disposition to the disease, prevailed throughout the whole of her crew. The lemon juice demanded from the sick and hurt board, had not yet arrived; and there was much hazard in this squadron going to sea, without a large supply of the vegetable acid. Having represented the state of the ships to Admiral Waldegrave, and his orders to go to sea not admitting of delay, under his authority
I pur-

I purchafed fourteen or fifteen chefts of lemons
and oranges, which were all that could be procured in Portfmouth.   Thefe fhips continued at
fea feven weeks ; but the fruit which was carried
from port was expended about the fifth week.
Fortunately however, they fell in with a fhip laden with lemons.   The Admiral ordered a quantity to be purchafed, fufficient for the fafety of
his people, during the remainder of his cruize.
They returned to port on the 7th of June, without lofing a fingle man in fcurvy.   The Admiral
was pleafed to order very particular accounts,
concerning the adminiftration and effect of the
lemons, to be tranfmitted to me.   Some very correct and well-detailed facts on the fubjects, were
in confequence fent : they will be found under
the article fcurvy *.——The weather during this
                                            cruize

* " *Minotaur, St. Helen's, June 9th,* 1795.
    " S i r,
    " After reading your treatife on naval hofpitals, I can have
little doubt of your making ufe of your utmoft exertions, to
promote any meafure that can contribute to the health and
comfort of our feamen.   It is from this confideration that I
now lay before you, copies of papers, the originals of which I
have juft tranfmitted to the Lords Commiffioners of Admiralty ;
not doubting, if feconded by your efforts on this occafion, but
that my ftatement will produce the defired effect.
    " Whilft I am on this fubject, I think it my duty to add,
that a few days after we began to ferve *grog* to the people, the
                        I 4   fcurvy

cruize, as mentioned by Mr. Sibbald, furgeon of
La Nymphe, " was remarkably fine ; there hap-
" pened to be, nearly during the whole cruize, a
" conftant

---

fcurvy made a moft rapid progrefs; and as this is by no means
the firft time that I have had occafion to make the fame obfervation, I am thoroughly convinced, that could it be made convenient to govenrment, to ferve out a wholefome genuine
wine (and fuch I apprehend might be found, without much
difficulty, or any confiderable expence) in *lieu* of fpirits, that
many lives might be faved, and the naval hofpitals lefs crowded with fcorbutic patients ; two-thirds of whom are loft to
the fervice, either by defertion, or the accumulation of frefh
difeafes,

I am, Sir, your moft obedient

humble fervant, .

To   (Signed)   Wm. Waldegrave."
Thomas Trotter, Efq. &c.

ANSWER.

Charon, June 10th, 1796.

" Sir,

" I am this day honoured with your obliging letter, inclofing Remarks, by the furgeons of your fquadron, on the effects
of lemons in the treatment of fcurvy.—What I have particularly obferved in thefe communications, is an increafe of the
fcorbutics, during the ufe of *grog.* This has been long known,
but never before fo clearly afcertained : but, befides confirm-
ing fome ideas whch I have entertained of the theory, it leads
to much valuable practical improvement *. I fhall make it

* Vid. Obfervations on Scurvy, fecond Edition, Longman,
London.

my

" conftant eaft wind, with heavy dews, every
" night."

April 17. The fquadron under Admiral Colpoys arrived at Spithead, after a month's cruize
in the Channel; during which, he had taken La
Gloire and La Gentile French frigates.

It has been mentioned, that this fquadron
was fupplied with lemon juice from the Charon,
from the idea I at that time entertained that they
muft foon need it. On their return to port the
fcurvy prevailed in every fhip. But, by confining the adminiftration of the lemon juice to the
worft cafes, no deaths happened at fea. The frigates, during this general fcorbutic taint, did not
feem to fuffer in proportion to the fhips of the
line : fome very bad cafes were, however, fent to
the hofpital from the Aftrea, in which fhip Lord
Henry

---

my bufinefs to put fome further queries to the furgeons, on
the fubject; and on drawing up a fummary of the method, and
fuccefs of treatment, of this fingular difeafe, in the fleet, for
the information of the Lords Commiffioners of Admiralty, I
fhall not fail to enforce fome regulations, which this occafion
has fuggefted ; not a little due to the humane attention which
you have paid to the health of the feamen in your fquadron.

I have the honour to be, SIR,
Your moft obedient
humble Servant,

To   ·   T. TROTTER.

The Hon. Rear-Admiral WALDEGRAVE.

Henry Poulett had fo handfomely diftinguifhed himfelf in the action with one of the frigates.

The Hannibal, one of Admiral Colpoys fquadron, put into Plymouth, over-run with fcurvy. This fhip had not been long in commiffion ; many of her crew were raw landmen, confequently more liable to be affected. After landing forty men at Plymouth hofpital, Captain Markham did not find them recover fo faft as he wifhed ; and feeing that they got few or no vegetables, owing to fome difpute with the gardener that had the contract, from pure compaffion to their diftrefs, this generous officer went to market, and purchafed them abundance at his own expence. This prevented the remainder of the Hannibal's fcorbutics from being fent to the hofpital : they were either cured on board, or fent on fhore in parties, to the different farm houfes in the neighbourhood, where the country people allowed them to take herbs and vegetables from their gardens, till they recovered their health.

Another fquadron, of four fail of the line, arrived from the North Sea, under the command of Rear-Admiral Harvey. Thefe fhips left Spithead a few days after the arrival of the fleet from the winter cruize. They fared ftill worfe than thofe under Admiral Colpoys. The Prince of Wales landed fifty men at Deal hofpital, five of whom perifhed in the boat; and fhe brought a
number

number ill to Spithead. The Thunderer, one of the ships of this squadron, returned in perfect health: her crew had lately shared prize-money for the St. Jago register ship, captured when they were in the Edgar; by this means they had furnished themselves amply, for the cruize, with every delicacy, even to live stock. The Russel was one of Admiral Harvey's ships, and was affected like the Prince of Wales.

April 20. No effectual change in the victualling department having yet taken place in the fleet, I was under the necessity of making further statements on the business, and thought it best to apply through the Port Admiral, to the Lords Commissioners of Admiralty; Earl Howe, though still at Portsmouth, and his flag down, was so indisposed from gout, that he was no longer in the official situation of receiving reports. I had nothing new to relate more than what has already appeared: the disease was advancing rapidly; the hospital was filling with scorbutics, from the ships at Spithead as well as those that had returned from sea, and if assistance did not soon arrive, there was danger that the whole fleet might be rendered inactive *.

Their

---

* The directions which the Admiralty had given, in some parts, were not followed with that dispatch which our accumulating danger claimed. I was therefore induced to write

to

Their Lordſhips were pleaſed to order the freſh beef to be ſerved to the fleet, in the uſual quantity ; and the Commiſſioners of ſick and hurt informed me, by letter dated the 29th of April, that they had ſent off, on the 27th, fifteen cheſts of lemons, and the like number of oranges, for Portſmouth, to be diſtributed to the ſhips afflicted with ſcurvy, as I might think proper. They added, that more ſhould be ſent as ſoon as demanded.

· Our wiſhes were now gratified. The ſurgeons no longer thought of ſending their men on ſhore ; they found the cures quicker accompliſhed under their own care.

At

---

to Sir Peter Parker a ſecond time. After waiting ſome days for their Lordſhips anſwer, Sir Peter's ſecretary told me, that the contractors *could not* get freſh beef, on which account the Admiral had not *thought proper* to tranſmit my letter to their Lordſhips. On informing Earl Howe of this buſineſs, he thought very differently, and by the firſt poſt ſent my letter to the Admiralty. It is neceſſary to ſtate all theſe facts as they happened. There is but one line of conduct to regulate the duty of an officer, *to be faithful to his poſt* ; and if misfortunes follow, he cannot be blamed. It will not be diſputed by any one, who conſults the authorities here produced, that the Channel Fleet was kept active by the meaſures I was purſuing, and the judicious practice of the ſurgeons, during this general and alarming calamity. My ſucceſſors, in a ſimilar ſituation, will benefit from the trouble which fell to my ſhare.

X

At the fame time the refident Commiffioner of the fick and hurt, received orders from that board, to furnifh the fcorbutics of the fleet *with as much fallad as could be procured.*

May 2. The Leviathan, Hannibal, and Swift-fure, dropped down to St. Helen's, as it was ex-pected, to accompany the Weft India convoy to a certain latitude, and return. The Hannibal, we have faid before, put into Plymouth in great dif-trefs, from a general fcurvy. But the ufual allowance of beef not being altered at that port, and Capt. Markham having provided his people with abundance of vegetables, his crew were well recovered. This was not the cafe with the Le-viathan; I therefore thought proper to fupply them with lemons, and the expreffed juice, in quantity for the cure of the difeafe, during the fuppofed fhort departure. They had, however, fealed orders to proceed to the Weft Indies; and after a tolerable paffage to Cape Nichola Mole, in St. Domingo, they did not lofe a man. Many cures were effected by the lemon; but a general taint now prevailed in all thefe fhips, to a dan-gerous degree, which had been gaining ground all the paffage. So fmall a fpace of the country being in poffeffion of the Britifh troops round the Mole, that relief for the fhips could not be obtained in that diftrict. The active humanity

of

of Captain Duckworth, however, foon found refources. He freighted a fmall veffel to the ifland
of Cuba, and received fruit and vegetables, from
which his people were reftored to health.

In this month the fcurvy appeared in the Royal
George, to the number of thirty cafes; and Mr.
Murray informed me, that a few were now feized
in the Queen Charlotte, and a general difpofition
towards it in all. Mr. Browne, of the Queen,
tells me in his report for this month, that a general
tendency to fcurvy prevailed among the crew.
There is no faying what were the reafons that re-
tarded the approach of the difcafe in thefe fhips.
The only circumftance of note is, that they had
not been at fea while the others had. *

May 26. The Royal Sovereign, Mars, Triumph,
Bellerophon, Brunfwick, with the Phæton and
Pallas frigates, failed under the command of
Vice-admiral Cornwallis. The Mars, a new commiffioned fhip from the Nore, was the only one in
this fquadron free from fcurvy when they went
to fea. The others had fhared of the lemons,
and

* " Many of the fhip's company are flightly affected with
fcurvy, but not fo far as to prevent them from doing duty: the
above number (twenty-four in fcurvy) is only fuch as are incapable of doing duty from that difeafe.

(Signed)

Sans Parcil, May 1796.   CHARLES BOWCARD, Surgeon."

and what fallad was fent us.   But as the general taint among the crews had not been corrected, I fupplied them from the Charon with a large allowance of fruit, and a quantity of juice in kegs. In the Triumph alone, as appears from Mr. Carthy's report for the month of June, one hundred and one cures were accomplifhed at fea! The condition of the Royal Sovereign and Brunfwick, was not much better than the Triumph.

This cruize was memorable, for the famous retreat of the fmall fquadron, on the 19th of June, from the enemy's fleet of twenty-nine fail; by courage and feamanfhip, that held up the naval character of Britain to the higheft degree of admiration.   It is only rivalled in the annals of the world, by a fimilar manœuvre, during the late war, under the fame officer; or to go farther back, it may be juftly compared to the retreat of the ten thoufand under Xenophon!

Admiral Cornwallis came to Plymouth on the 24th of June, to refit.   He failed early in July, to relieve fome of Lord Bridport's fhips, off Bellifle.

During the month of May, all the fhips which returned before, from different cruizes, to Spithead, were ftill increafing in the number of fcorbutics. The London had, according to Mr. Smith's report,

port, ninety-nine under cure. * The Valiant, according to Dr. Thompfon, ftill more; and the Coloffus, according to Mr. Ballentyne's, was fixty-eight. In the Barfleur, fixty; Prince of Wales, eighty; Robuft, fifty; Ruffel, fixty-four; and the others more or lefs affected.

The fallad, which I had reprefented in my letter to the Admiralty, as the beft vegetables for our relief, was fupplied in too fmall quantity to eradicate the difeafe. I wanted it to be given in large allowance to the different meffes of the feamen; and that the frefh beef broth fhould be full of greens, or other pot-herbs. The lemons we could only confider for the cure. I demanded, therefore, five thoufand pounds for daily confumption : and walked over the markets and gardens, to inform myfelf if this could be obtained. Our fupply, hitherto, had feldom exceeded *one hundred pounds* of lettuces, young onions, and muftard. †

Admiral

---

* " The fcurvy is ftill prevalent among the fhip's company, in a violent degree. There is fcarce a day but two or three are added to the fick lift, and unfit for duty; but foon yielding to the lemons, oranges, and falladding, now fupplied.

(Signed)

London, May 1795. JOHN SMITH, Surgeon."

---

† The reader may fmile at the idea of a Phyfician to a Fleet, attending the ftalls at a vegetable market, or preambulating

Admiral Lord Bridport came to Portfmouth about this time, to take the command of fourteen fail of the line. He had heard of the general fcurvy, but fuppofed, from the orders given fome time ago, by the Admiralty, that it was effectually overcome. His Lordfhip received my opinion on the fubject with much attention, and entered earneftly into my manner of curing it. He defired the refident Commiffioner of fick and wounded, to order the quantity which I deemed neceffary, whether it could be procured at Portfmouth, Gofport, or elfewhere. It was of little moment whether the fallad ought to be confidered as a part of victualling, or medicine; the public fervice demanded inftant relief. Had it been in my power to command it, it fhould have been brought from the Land's-end, in Cornwall, before the fleet had fo long groaned under the affliction. We have heard of a Minifter * ordering a train of ordnance acrofs the country, from Woolwich to Portfmouth, to fave time; in this manner would I have wheeled the product of every diftant garden to Portfmouth, left the tooth of a failor

* The firft Lord Chatham.

K   fhould

---

bulating the country, to calculate the produce: but it never appeared to me below the dignity of the profeffion; nor did I confider it a mean tafk to ferve the fallad with my own hands, from the Charon's quarter-deck. *

ſhould drop from his gums, by a tardy conveyance of his deliverer. *

Deliberat Roma, perit Sagantum.

June 1. We now received upwards of four thouſand weight of ſallad, daily. It was given in greateſt quantity to the ſhips moſt affected : theſe were, the London, Prince of Wales, Valiant, Coloſſus,

* TO THE COMMISSIONERS OF SICK AND WOUNDED SEAMEN.

Spithead, May 24, 1795.

GENTLEMEN,

I am duly honoured with your letter of the 22d inſtant : And beg leave to remark, that we, on the ſpot, are by no means of opinion that any ſupply of vegetables, yet ordered for the uſe of the Fleet, is likely to prevent the progreſs of ſcurvy among the ſeamen. For five days paſt, ſomething above one hundred weight of ſallad has been ſent by Dr. Johnſtone to the ſcorbutics, and diſtributed at the rate of *four oz. per man.* It appears that ſeven hundred caſes have been cured by the fruit alone ; not including the cures in ſhips at ſea ; and as freſh caſes occur, they will be eaſily overcome, while we poſſeſs the lemon and orange. But means of prevention muſt be extended to *every man,* otherwiſe we can promiſe no ſecurity. The ſquadron of Rear-Admiral Colpoys returned to port with three hundred ill of ſcurvy. The Prince of Wales and Ruſſel, under Rear-Admiral Harvey, have been equal ſufferers ; and had not the ſhip' under Rear-Admiral Waldegrave, received fourteen thouſand lemons and oranges, when they ſailed, their ſituation muſt have been ſtill worſe. For theſe reaſons I have ſent to the ſquadron about to ſail under Vice-Admiral Cornwallis, what fruit we could ſpare.

The

Coloſſus, Robuſt, and Ruſſel. But the Royal
George was becoming as bad as any of them, according to Dr. Higgin's report; ſhe was therefore conſidered, in allowance, with the others.

June 10. Since the large allowance of ſallad was
ſerved, the ſcurvy has continued to decline. The
ſudden change is wonderful. The ſquadron being

K 2   - now

The gardens in this neighbourhood, I have found from my
own inquiries, can ſupply vegetables, equal to our wants,
at a very moderate price; young onions and raddiſhes have
been ſold at this market for a halfpenny per bunch, for
the laſt three weeks; theſe, with what lettuces, &c. may
be procured, and ſerved in certain proportions, to the
meſſes of ſeamen, I am apt to think, would effectually correct this ſcorbutic taint in the ſpace of a fortnight. Not
leſs than ten or twelve cart-load of vegetables were yeſterday morning brought to Portſmouth market, beſides the
quantity ſent to Portſea, Goſport, and what might be got
ſtill cheaper in the Iſle of Wight.

The Channel Fleet, being now ordered to aſſemble, I
conſider it my duty to give you this information, leſt the
Lords Commiſſioners of Admiralty ſhould expect our afflictions to be remedied through the directions of your
Board.

I have the honour to be,

GENTLEMEN,

Your moſt obedient,

and very humble ſervant,

T. TROTTER.

now ordered to fea, the health of the men was confidered equal to the expected cruize ; but it could not be fuppofed fecure againft the effects of a fea diet, where vegetables and frefh beef had no fhare. The Surgeons were ferved from the Charon with boxes of lemons, in proportion to the progrefs of fcurvy lately marked among the people, and thirty gallons of lemon-juice, to be ufed when the fruit was expended. I referved two hundred and fifty gallons in the hofpital fhip, left any un-forefeen length of time, or exigencies of fervice, might keep the fleet at fea beyond what we forefaw.

I will give the condition of the Coloffus, in the beginning of this cruize, as an example of the ftate of the fleet : the Surgeons of the whole having confirmed what Mr. Ballentyne fo emphatically expreffes in his remarks. It appears that Mr. Ballentyne cured no fewer than one hundred and twenty, from the 1ft to the 12th of June *, while the large allowance of fallad was iffued.

*, " State

✻ " State of HEALTH, on board his Majeſty's Ship Coloſſus,
from the 1ſt to the 30th of June, 1795.

DISEASES.	Since laſt Report,				Preſent Sick Liſt.
	Taken ill.	Sent to the Hoſpital.	Recover-ed.	Dead.	
Fever - - -	3	2	—	—	1
Flux - - -	1	—	—	—	1
Catarrhal Complaints	6	—	9	—	1
Scurvy - -	60	—	120		
Ulcers - - -	1	—	—	—	3
Venereal Complaints	4	—	2	—	4
Wounded Men, &c. June 23.	53	2	29	—	34
Total - - -	128	4	161	—	44

REMARKS.

" The above ſtatement is a proof of the good effects of
fruit and vegetables in the cure, as well as the prevention of
the ſcurvy.   I have not had a patient in the liſt, for ſcorbutic
ſymptoms, ſince the fleet left Spithead.

(Signed)

Coloſſus, at Sea, June 30, 1795.   J. BALLENTYNE, Surgeon."

K 3

During

During the prevalence of this general difeafe in the fleet, after the liberal allowance of fruit came, and the quantity of fallad increafed, no feaman was known to exprefs the leaft regret at being kept on board for the cure: on the contrary, they were not a little delighted with the novelty of the bufinefs, and refinement in diet from an Admiralty order: the whole being ferved from the Charon, fhe was called the *Doctor's garden.*

From the middle of March to the 12th of June, upon comparing notes and remarks from the reports of furgeons, it appears that not lefs than three thoufand cafes, unfit for duty, had been cured, on board, by the fruit, or the preferved juice. About twice that number, with flighter fymptoms, were relieved by the fruit, the juice, and the fallad. Thus were thofe men kept ready for fervice, our fhips companies preferved entire, and defertion, and bad habits often contracted at hofpitals, prevented.

June 12. This day the fquadron under Admiral Lord Bridport failed; confifting of the following fhips: Royal George, Queen Charlotte, Queen, Prince of Wales, London, Prince George, Prince, Barfleur, Sans Pareil, Valiant, Coloffus, Irrefiftible, Ruffel, Orion, Frigates, &c. with the Charon hofpital fhip.

June 23.

June 23. The fquadron captured three fail of the enemy's line, off the Ifle of Groa, with very little lofs of men on our fide.

SHIPS TAKEN.

Le Tigre - - - 74 guns.
L'Alexander - 74
Le Formidable 74

The fervice of our fquadron, it now appeared, was to protect the expedition under the command of Sir J. B. Warren, againft Quiberon. The Peninfula was foon taken poffeffion of; but quickly retaken by the French, and the emigrant troops that were landed, to the number of five thoufand men, were either put to the fword, or put to death by military execution, afterwards!

After the victory on the 1ft of June, 1794, as mentioned before, many of our fhips received a contagious fever from the French prifoners. On this occafion I did not find that any particular ficknefs prevailed in the captured fhips; certainly nothing of the kind was communicated to our people.

In the early part of the cruize, a low fever appeared in the Prince. It was not attended with much mortality, nor was there any thing peculiar in its nature. Mr. Folds treated it with antimonials, in the firft ftage, and then gave bark, wine, and opium.

K 4   July 1.

July 1. Two men in fmall-pox, were fent to the hofpital fhip, from the Orion. The man firft affected, was convalefcent, and had been near three weeks ill on board, by which means others were infected. The infection was brought to the Orion by a Quarter Mafter, from Haflar Hofpital, at the time the fhip failed from Spithead.

July 6. This day, eleven men, ill of typhus fever, and ten in the fmall-pox, were received from the Orion. Two of the fever patients, had been in a ftate of delirium from the beginning; and the whole of them had a very bad appearance. This fever evidently fpread from contagion. I wrote to Sir James Sumaurez on the fubject; who requefted me to come on board the Orion. Upon having the whole of the fick, who were able to walk, upon deck, I found feveral in a ftate of convalefcence, that had been ill of the fever. They were men lately entered, for the bounty given by the counties, according to the late act of Parliament. The infection had been acquired on board the Royal William receiving-fhip, at Spithead; they told me, in the prefence of their officers, that they conceived their difeafes to have been caufed by fleeping on the deck, the crowded ftate of that fhip not allowing them room to hang up a hammock.

Every perfon on board with the fever, being now moved to the Charon, Sir James Sumaurez ordered

ordered a ftrict look out to be kept at others that might complain: and the Admiral gave inftructions that the Orion fhould ftay near the hofpital fhip, that the infected might be feparated on the firft fymptoms of indifpofition.

Sir James Sumaurez had taken the command of the Orion, with a raw fhip's company, only a day before we went to fea. The people's cloaths were now ordered to be aired abroad with their beds; their perfons to be cleaned, and flops ferved to thofe in want. The fhip was pure and clean in every corner; the common procefs of fires, &c. below, was attended to; and no frefh cafe of fever appeared afterwards.

Some of thefe patients had fymptoms of confiderable malignity; and three of the number died on the fifth day from being feized. No nurfe or attendant in our hofpital was infected: I attributed this folely to the extreme exact attention to cleanlinefs, both in the perfons of the fick, their body, linen, and bedding: a boiler with water was kept in conftant readinefs to wafh every article of cloathing as foon as it came from the beds of the fick, and a tub ftood by for this purpofe. Not even a handkerchief or night-cap was laid away till wafhed and aired.

Only one patient died in the fmall-pox; which were of a milder kind than ufually met with in
adult

adult fubjects. As it was likely that the difeafe muft now extend itfelf to every one on board the Orion, who had never been infected, people of this defcription, to the number of fifteen, were called to the quarter deck, and admonifhed by the captain and myfelf, to fubmit to inoculation, as an eafy and fafe mode of getting over the difeafe. Some of their prejudices, however, were not eafily overcome; they were of the religious kind, and they did not confider it right to bring a diforder upon themfelves. We combated this objection with the ufual arguments, that Providence had put into our power the means of efcaping a dreadful diftemper by a trifling operation, and that it was impious in human beings to neglect it. They felt our advice more fenfibly when they were told, that we confidered it our duty to inftruct them for their welfare, and that our only motive was their fafety, for they were not to be compelled to undergo inoculation; but act as they pleafed. I added, that two or three general inoculations of feamen had taken place on board the Charon, not one of whom had ever been confined an hour to bed. Ten of the fifteen confented to be inoculated, and had the difeafe in its mildeft degree; the other five were doubtful of having had it in their infancy, and were not infected. *

* At this time four men and boys were inoculated in the Charon, and did well.

July

July 9th. This day the wounded men of the Fleet, to the number of forty-five, were moved from their own fhips to the Charon ; among them was Captain Grindall, of the Irrefiftible, who was feverely wounded in the breaft, and fhoulder, and arm.—This evening we parted with the Fleet, to proceed to England, under convoy of the Orion.

On this morning the fquadron under Admiral Cornwallis joined Lord Bridport ; confifting of the Royal Sovereign, Formidable, Triumph, Bellerophon, Invincible, and Brunfwick : in confequence of this relief, the Prince of Wales, Queen Charlotte, Ruffel, and Coloffus, returned home : the Prince and Barfleur parted a few days before with the prizes.

15th. This day put into Weymouth, and landed Captain Grindall at his own houfe ; the harbour was lined by the inhabitants, and the Captain was heartily cheered on fhore.

We purchafed ftock, fruit, and other refrefhments for the fick and wounded.

16th. Arrived at Spithead, and landed the fick and wounded at Haflar.

The weather, during the early part of this cruize, was cold, with eafterly winds ; but very fine at the time we left the Fleet.

It

It appeared that a feaman belonging to the Queen Charlotte, who affifted the wounded men from the boat into the lower deck of the hofpital fhip, at fea, was foon after feized with the fmall-pox. The Orion's people lay near the port where he entered. The eruption appeared on this man the day after the Queen Charlotte arrived at Spithead: he was fent on fhore immediately, and the difeafe fpread no farther.

Auguft 1ft. The Charon was ordered to return to the Fleet. We were now fupplied with a quantity of ftock, vegetables, porter, &c. to be diftributed to the fhips for the ufe of the fick; articles which the length of the cruize, and the former fufferings of the feamen in fcurvy, appeared to me to render neceffary at fea.

Auguft 5th. The Charon failed under the protection of the Thunderer.

Auguft 15th. Joined the Fleet off Groa.— This day I was employed in vifiting the fhips, and diftributing the ftock and other refrefhments from the Charon.

State

State of HEALTH in his Majefty's Ships under
the Command of ADMIRAL LORD BRIDPORT,
Auguft 17, 1795.

SHIPS.	Number on the Sick Lift.	PREVAILING DISEASES.
Royal George -	45	Scurvy.
Sans Pareil - -	36	Scurvy, increafing rapidly.
Invincible - -	52	A low Fever and Scurvy. *
Valiant - - -	35	Obftinate Venereal Complaints, combined with Scurvy.
Formidable - -	22	Bad ulcers: Scurvy beginning.
Irrefiftible - -	35	Scurvy, increafing faft.
London - - -	41	Slight Scurvies.
Brunfwick - -	30	Scurvy, increafing.
Royal Sovereign	30	Inveterate Scurvy.—Eighty with fymptoms, but doing duty.
Queen - - -	43	Scurvy and obftinate Venereal Complaints.
Triumph - -	29	None.
Prince George -	53	Scurvy, increafing faft.—Some inveterate cafes.
Bellerophon - -	24	Scurvy, beginning.

P. S. The Scurvy is alfo appearing in the Frigates, particularly in La Nymphe and the Thalia.

* Vide Mr. Kenning's Remarks on this Fever, under the article Typhus.

In the above report to Lord Bridport, the principal caufe for animadverfion was the re-appearance of the fcurvy. It was only for the fourteen

teen days preceding our return to the Fleet, that
the difeafe had fhown itfelf. The weather had
for fix weeks paft been mild and warm. But the
ftation of the Fleet, which was off Belleifle, or
always within a few leagues of it, left the minds
of the feamen in a ftate of inaction. The fate
of the emigrants that were flaughtered on the
peninfula of Quiberon, and the fubfequent execution of the remainder, wrung every foul with
pity and horror. I think it neceffary to mention
all thefe circumftances, as they might have a
fhare in difpofing the habits of our people to the
difeafe in queftion.

It was now that the furgeons felt the value of
the repeated fupplies of lemon juice ; and in this
trip to the Fleet we had renewed our ftock of it.
The fhips that were in want of other neceffaries
for the fick, received them from the Charon, and
I recommended every where a fupper of fowens,
with fugar and wine, for the fcorbutics. The
mutton and vegetables were made into broth,
and while they lafted our people fared fumptuoufly.

Although the fcurvy had been checked from
the large quantity of fallad ferved for the laft ten
days at Spithead, yet we were not fo fecure as
to think it would not return during a long cruize.
It is to be remembered, that moft of the fhips
had now been ten weeks at fea. Had the fallad

7   been

been given fooner, or continued to be ferved longer, we fhould not have had fo quick a return of the difeafe.   There was alfo, at this time, on my part, fome reafon for regret, that the improvements in diet, which I had fuggefted fo early as the month of March, had not been complied with.   Even the molaffes were neglected to be fent, when we firft failed ; they would have made the oatmeal gruel palatable, and it is a grofs miftake to think that feamen will take it without this addition : it is even cruel to think that they ought.   There was another circumftance to be confidered, the water was now growing fcarce, and as there feemed no profpect of the fhips being foon relieved, it was put to allowance.

20th. In this fituation the fick lifts of the different fhips continued to vary and fluctuate ; as a few fcorbutics were cured, an increafed number came into the lift.   The lemon juice retained its powers, and was every where effectual.   Mr. Moffat, furgeon of the Aquilon, mentions in his report, that what he had kept for a length of time ftill cured the difeafe.   This was part of the fupply granted in September 1794 ; fo that it had been near two years expreffed from the fruit.*

This

---

* " The fcorbutic complaints have uniformly yielded to lemon juice, notwithftanding its having been on board for near fifteen months, and, confequently, confiderably impaired in its virtues.
(Signed)

Aquilon, July 24.   THOMAS MOFFAT, Surgeon."

This is an incontrovertible fact, that this valuable article may be preserved at sea for any voyage, and secure health to a ship's company, when every thing else shall fail.

The sick and infirm of the ships in the squadron were sent on board the Charon. They are chiefly chronic complaints. Eight cases of bad ulcers in the leg were sent from the Formidable, with a scorbutic appearance. No other case of scurvy has been received.

August 29th. The scurvy is still gaining ground. The remainder of our lemon juice has been distributed to the ships, and a hundred gallons, in reserve, are sent on board the Royal George. The London, and Prince George, have been dispatched to Plymouth, their water being nearly out.

Mr. Milligan, surgeon of the Megœra fireship, being moved into the Charon, the treatment of a number of scorbutics devolved on Captain Dickson and his officers. The disease went over the whole ship's company. But the cure of scurvy was now grown so familiar to our officers, that Captain Dickson prescribed the lemon juice to his people with the usual success, and recovered many.

This evening the Charon and Crescent frigate parted from the Fleet for England.

September 3d. Arrived at Spithead, and landed the sick at Haslar.

The

The Robuft, which, fhip had failed in June with Sir J. B. Warren, remained for three months with that fquadron in Quiberon Bay, and was now at Spithead.   As long as the fruit and lemon juice were ferved, the fcurvy was kept under; but thefe being done in July, fhe landed fixty-nine at Haflar, in the laft ftage of the difeafe.   Three died before the arrival in port*.

Admiral Harvey having failed on the 31ft of Auguft, with the Prince of Wales, Queen Charlotte, Prince, Orion, Mars, Minotaur, Ruffel, Thefeus, and Tremendous, to relieve the fquadron under Lord Bridport, on the 19th of September his Lordfhip came to Spithead.

September 20th.   The fhips of this fquadron had on board, at their arrival, the following number of fcorbutic patients; who were reported to
me

* " Two of the patients that died in fcurvy had lately been much reduced by the flux; the other complained only two days before he died; which, I confefs, was then rather un-expected.   The cafk of lemon juice you were fo kind to give me, on the day we failed from Spithead, was expended laft month.   The Captain Thornborough gave me a cafk of porter, which I thought was ferviceable; but notwithftanding that, and every thing I could poffibly procure, had we remained out ten days longer, our condition would have been truly diftreffing, as moft of the fcorbutic patients were in the *laft ftage* of that dreadful difeafe.

(Signed) JAMES TURKINGTON, Surgeon."
Robuft, Aug. 1795.

L

me by the surgeons, or surgeons' mates, at a general visit.

Royal George - - -	160
Royal Sovereign - - -	250
Queen - - - - -	78
Sans Pareil - - - -	100
Invincible - - - - -	260
Valiant - - - - -	100*
Triumph - - - - -	30
Bellerophon - - - -	30
Pallas - - - - - -	17
Megœra - - - - -	60
	1,085

The Formidable put into Plymouth. Of the number mentioned above, one hundred and seventy of the worst were sent to the hospital.

September 22d. The Commissioners of sick and wounded having informed me at this time, when fitting the Charon for another trip to Quiberon, that their stock of lemon juice was low, and no more to be procured, I directed the surgeon of the Charon to purchase fifty bushels of apples for the use of Lord Bridport's ships. The Royal

---

* Dr. Thompson assures me, that not one of the Valiant's people that complained of scurvy during the late cruize, were among the number who had been ill at Spithead in April and May, and shared so abundantly of the lemons and sallad.

Sovereign

Sovereign having fent no men to the hofpital, a large allowance was given to that fhip.   An extra allowance of vegetables was put into the frefh beef broth.

September 29th. The quantity of apples was repeated.   Even at this time there are fcarce any vifible remains of fcurvy in the Fleet.   The apples were very acceptable; the more fo, becaufe every failor had been long ufed to the lemon juice.

The Prince George arrived from Plymouth laft week, in perfect health, to receive the flag of Rear-Admiral Chriftian, intended for the Weft Indies.   Mr. Harris, the furgeon, informs me, that he had cured all his fcorbutics on board, but four that were in a bad condition.   This he was enabled to do through the bounty of Mr. Unwin, the purfer, who had fupplied the fhip with vegetables in great abundance.   This gentleman reafoned like a man of feeling.   ' He was in the receipt of a handfome income, he faid, by victualling the fhip's company, and thought he could not act juftly if he forgot them in their diftrefs.'   May this example, in fimilar fituations, be imitated ! Mr. Sheppard, purfer of the Vengeance, fupplied the fcorbutics of that fhip in this manner; by which means I was enabled to cure them all on board *.

* Vide Medical and Chemical Effays; Jordan, London, 1795.

The

The London continued inactive at Plymouth; (her purſer being only acting for the time, had it not in his power to increaſe the vegetables;) and landed upwards of a hundred men at the hoſpital, ſome of whom were never returned. For ſimilar reaſons, the Robuſt ſailed on the end of this month to join Admiral Harvey, and left ſixty of her ſcorbutic people at Haſlar.

September 30th. The Charon being again ordered to proceed to the ſquadron off Belleiſle, a quantity of ſtock, apples, vegetables, porter, &c. has been received on board, for the uſe of the ſick in their own ſhips, as well as in our hoſpital.

In the mean time the Lords Commiſſioners of Admiralty having obſerved the good effects of ſupplying the ſeamen with refreſhments, had ordered bullocks and ſheep to be ſent from Cork and Plymouth, with potatoes, cabbages, and onions, for Admiral Harvey's ſhips.

September 30th. This day Earl Spencer, accompanied by Lord H. Seymour and Mr. Pybus, lords of admiralty, honoured the Charon hoſpital ſhip with a viſit.

Next day their Lordſhips ſurveyed Haſlar Hoſpital, and marked out the ground for erecting the houſes for a governor, lieutenants, and other officers; about to be added to this inſtitution, in conſequence of a general inquiry, by two flag officers

and

and two captains of the Fleet, made in March 1794, at my reprefentation.

October 3d. The hofpital fhip failed from St. Helen's.

October 6th. Joined the fquadron under Rear-Admiral Harvey off Belleifle; and diftributed the ftock, fruit, and vegetables, &c. among the fhips.

The fhips were all healthy.—The weather hitherto had been remarkably fine. During a fquall of thunder and rain a few days before, the lightning had fhattered the mizen-maft of the Ruffel, and killed the firft lieutenant, while fitting at the wardroom table. She went to England immediately.

Received a man from the Marlborough in fmall-pox: care had been taken to feparate him from the fhip's company, and the difeafe, Mr. Kent informed me, fpread no farther.

October 9th. The fquadron anchored in Quiberon bay, between the little iflands Hedic and Houat, during a hard gale of wind and thick weather from the weftward. In this bay we found part of Sir John Borlafe Warren's fquadron, and a large fleet of tranfports and victuallers; the other frigates had failed on an expedition againft Ifle Dieu.

October 18th. Received a patient in small-pox from the Queen Charlotte. This ship had now been nine weeks at sea, and there could be no suspicion of the infection having been received directly from any person under the disease. Mr. Murray, therefore, accounted for it in this manner: The husband of one of our nurses was a soldier on board the Queen Charlotte: and while the patient from the Marlborough in small-pox lay in the hospital, this woman had attended him, and carried the infection when she visited her husband about a week before the man sickened. The disease was of the confluent kind; he died on the fourth day. This fact ought to make us very circumspect wherever small-pox or other contagious diseases are raging.

The sick of the squadron, to the number of fifty, were sent to the hospital ship; which included a few from the ships under Sir J. B. Warren. They are chiefly diseases of the season, rheumatisms, pectoral complaints, and some fluxes from the Standard. The seamen on board the transports have suffered from the flux: many have perished, as much from the want of diet and necessaries, as medical assistance. Some have been relieved from our hospital by the Admiral's order.

From this date to the 3d of November was almost a continued gale of wind.

§   Nov. 3.

Nov. 3. The wind coming to N. N. E. the squadron failed from Quiberon bay. Next morning the transports got under weigh, and followed us.

Nov. 7. Received the remainder of the sick from the ships; which makes the whole on board sixty-eight. Parted with Admiral Harvey, in company with two frigates and a fleet of transports, with troops.—Remnants of Pharsalia and mighty battles fought in vain!

Nov. 10. Put into Scilly, the wind blowing strong from the eastward.

Nov. 11. Mr. Millegan employed in procuring stock and vegetables for the sick in the hospital.

Nov. 12. Sailed from St. Mary's Sound, Scilly.

Nov. 16. Arrived at Spithead, and landed the sick at Haslar. The Charon was ordered to receive troops, and prepare for the West Indies, under R. A. Christian. The medical staff was ordered to another ship.

— 1796. —

Jan. 2. This day the squadron under Rear-Admiral Harvey arrived at Spithead, after an absence of eighteen weeks from England. On board the ships came passengers 3,000 soldiers, embarked from Isle Dieu, when evacuated.

This squadron had been repeatedly supplied with refreshments, from Plymouth and Cork,

L 4   while

while on the coaſt of France. The ſcurvy appeared in a number of caſes, in all the ſhips, but
was quickly cured by the lemon juice. Some
deaths happened from the dyſentery: this complaint probably originated from the ſcorbutic
habit, with which it is ſo often combined. But
ſome of the ſurgeons, not without reaſon, ſuſpected that it was cauſed by bad water obtained from
the ſmall iſlands in Quiberon Bay. The wet
weather might however produce it, with the cold
of the ſeaſon, without any other cauſe.

Jan. 10. The above ſquadron, with five ſail of
the line lately returned from ſea, under Vice-
Admiral Cornwallis, were ordered to be ſup-
plied with vegetables, beſides the allowance of the
purſers.

Feb. 1. No ſickneſs of any kind in the ſhips
of the Fleet.

The winter, hitherto, has been free of froſt;
but gales of wind from S. W. with rain ſtill continue.

April. A general order has been given, from
Admiralty, in conſequence of a recent application
from me, to ſupply every ſhip and ſquadron of
the Fleet with the under-mentioned allowance of
vegetables, for the firſt fortnight after returning
from ſea on meat days; which is at the rate of
four days in the week, for mixing with the broth;
it has been recommended to diſtribute the onions

to

to the meffes, that the feamen may ufe them as they pleafe ;

For every hundred men of the complement, and in proportion,

    Cabbages or greens, in feafon - 50 lbs.

    Onions   -   -   -   -   10 lbs.

N. B. Other vegetables are to be fubftituted by the Agent Victualler, when the above mentioned, cannot be procured.

April 14. An eafterly wind has prevailed for fome weeks : the weather cold, but not productive of any difeafe in our fhips.

The duty of the Fleet being now done by fmall fquadrons, and from the liberal allowance of vegetables to the feamen on returning from a cruize, we may date the extinction of the fcurvy.

June 26. This day I was ordered by Vice-Admiral Colpoys, to vifit the Niger frigate, where a contagious fever had lately made its appearance.

It firft appeared among the marines, but latterly affected the feamen. Its origin could not be exactly traced; but probably it was brought from a fmall veffel the Niger had captured on the coaft of France, which had juft landed fome foldiers at Bayonne from Breft.

In fourteen days from the time this fever was firft detected, forty-four people were fent on fhore. This fhip being under excellent difci-

plinc,

pline, I did not defpair of ·feeing it quickly fubdued; and the cafes that were fent for the laft
four days were very flight indeed *. None died.

July

* "This fever made its appearance by vomiting and
head-ach, which in fome cafes was foon followed by delirium.
In others delirium was the firft fymptom.

June 19. Two marines complained: next morning they
were fent to the hofpital. The fhip was wafhed, aired with
windfails, ftoves placed in different places: cloathing and
bedding all expofed to the air, &c.

21. Two more marines fent to the hofpital. Fumigated with
tobacco.

22. White-wafhed the fhip, aired the cloathing, great part of
which, with part of the bedding, wafhed this morning.

23. Four men fent to the hofpital; two of them marines.
Fumigated with gun-powder.

24. Five men fent to the hofpital; four of them marines,
Bedding, &c. aired. Fumigated with tar. Wine ferved from
to-day till the 30th.

25. Eleven men fent to the hofpital, five of them marines.
Fumigated with tobacco.

26. Nine men fent to the hofpital. Fumigated with tobacco; cloathing and beds aired every day *.

27. Fumigated with charcoal and brimftone. No perfon
complained till after fumigation, when two men who went
immediately below after the tarpaulins were unlaid, complained of head-ach, one was delirious: they were immediately
fent on fhore.

———————

* Thefe nine men were examined by me; the degree of ftupor in fome of them was confiderable, with that peculiar look
which indicates fatuity or imbecility of mind. I did not think
any of them in danger. T. T.

28. Five

July 20. Mr. Burd furgeon of the Niger, informs me, that feven of his men returned from
Haflar, have relapfed.   The greater part of the
cloaths of thefe people having been fent on board
unwafhed

---

28. Five men complained in the night, very flightly affected.

29. Fumigated.

30. Five men fent to the hofpital.

July 1. Fumigated with charcoal in the morning, and nitrous gas in the evening.

2. Every man was ordered below ; the fcuttles clofe fhut,
tarpaulins laid over, &c. for half an hour, during which time,
the nitrous gas was diffipated through the fhip, according to
the directions given in Dr. Smyth's Pamphlet.   Very foon after it was begun to be ufed, a number of men were affected
with coughing ; and before the half-hour was expired, the
coughing became more general, and in many attended with
head-ach, which did not leave them till after walking the
deck in the free air for a confiderable time.

July 3. Doctor Smyth's procefs was repeated, and attended
with the fame effects as yefterday : which in my opinion,
from the irritation produced in the lungs, is a proof that it is
in fome degree unfavourable to refpiration.   When it was
ufed after the fumigation with charcoal and brimftone, it
fomewhat removed the difagreeable fmell produced by that
procefs."—*Extract from* Mr. BURD's *Letter, tranfmitted for the
information of Admiralty, by Capt. Foote.*

Niger, July 8th, 1796.

On the 27th, it was remarked that the attack of this fever
became fo flight as no longer to make it an object of much
attention

unwafhed from the hofpital, Captain Foote
thought it proper that they fhould be cleaned
immediately. The men therefore rofe early to
do this, and were foon after feized with convulfions and fever. It is doubtful with me whether
this relapfe ought to be attributed to fleeping
again in beds or cloaths not thoroughly purified; or
to the cold they might experience on rifing fo
foon. It is however a certain proof, that it is an
improper cuftom not to wafh every piece of
cloathing belonging to the feamen, while at an
hofpital.

Aug. 17. A fquadron of five fail of the line
arrived at Spithead, after a cruize of thirteen
weeks, off the Weftern Iflands, under the command

---

attention. The feafon of the year being favourable, I look
upon the fpeedy extinction to have been accomplifhed folely
by the timely feparation; the other methods practifed, are
nothing more than the routine of duty and cleanlinefs in a well-
ordered fhip.

The coughing and head-ach are the common effects of air
being refpired that is impure, and not fufficiently abounding
with *oxygene*; they were therefore owing to the deficient ftimulus of the atmofphere, and difappeared by going on deck,
where the common air had a larger portion of *vital air*. The
charcoal and fulphur were ordered to be tried by Dr. Johnftone, the refident Commiffioner of fick and hurt; but thefe
had no advantage over the gun-powder, which Captain Foote
ufed. They both give out fulphurous gas: but I confider them
both hurtful, not merely paffive, but in direct oppofition to
the purpofes of ventilation. See the article CONTAGION.

T. T.

mand of Rear-Admiral Lord H. Seymour. All
of thefe fhips fuffered more or lefs from fcurvy,
during their time at fea, although his Lordfhip
ordered them to be fupplied with fome refrefhments from the iflands. The Thefeus failed
with a difpofition to the difeafe among her people, owing to her being much in the Channel
before; and Mr. Snipe, with infinite trouble on
his part, kept it under by the citric acid, and a
quantity of pickles which he had previoufly prepared with his own hands in harbour. Twenty
four cafes unable to do duty occurred in the
Jufte, and were cured by Mr. Kenning in his
ufual way.

The taint of fcurvy prevailed ftill more in the
Triumph. She had been much at fea in the early
part of the fpring; and Mr. Moffat in his report,
remarks, that " the fcorbutic cafes, as ufual, yield-
" ed to the lemon juice in every inftance; and
" I may venture to affert, that had there not
" been fo plentiful a fupply of it, one third of
" the fhip's company, at our return would have
" been laid up with fcurvy. So general was the
" tendency to it, that almoft every cafe of con-
" tufion or ulceration was attacked with it, nor
" could their cure be accomplifhed without a
" few dofes of the acid. The number ftated
" above, (forty-fix) are confequently not the whole
" who had taken the acid, but only fuch cafes
" as were unconnected with other complaints
                                    " and

" and unable to do duty." Mr. Moffat alſo ob-
ſerves that ſome rheumatiſms inſtead of being relieved by the warm latitudes, were rather aggravated by them. It is probable that the ſcurvy
had much to do in theſe caſes, for the diſtinc-
tion is often difficult; a trial of the citric acid
would have decided the point.

Some deaths happened during this long cruize;
but they were the common diſeaſes of human
nature, and not to be imputed to any thing peculiar to a' ſea life *. Mr. Nepecher reports the
Orion to be perfectly well.

Sept. 1. Some very warm weather prevailed
towards the end of laſt month. The ſhips lately
returned from ſea, have been effectually recruited
by a large allowance of freſh vegetables.

Sept.

---

* " The death that appears in the report for June, occaſioned by hydrothorax, was rather ſudden. The patient appeared to have a complicated diſeaſe, perhaps with ſcurvy. For a
few days he took the citric acid, without relief: afterwards,
large doſes of æther and tinct. opii occaſionally relieved his
breathing: he died on the 7th day after he complained. On
diſſection there was found in the right cavity of the thorax, a
large quantity of very offenſive white coloured fluid: the lobe
of the lungs in that cavity appeared ſmall and ſhrivelled up to
the ſuperior part of the cavity. In the left ſide, the lung was
all diſſolved to a ſmall piece of ſoft membranous ſubſtance;
the quantity of fluid was not leſs than four quarts. The heart
and veſſels had their natural healthy appearance; and the
common quantity of fluid in the pericardium: the abdominal
viſcera were in good order. It was really extraordinary that
the

Sept. 20. Arrived at Spithead, the Queen Charlotte and London, under Admiral Colpoys, from Plymouth, the other ships having failed to join Admiral Gardner off Ushant. Mr. Smith in his report mentions the death of Lieutenant Bell, of a violent inflamation of the lower part of the abdomen, that resisted all remedies. The scurvy for the last two months appeared in twenty-seven cases, which were cured on board the London, with Mr. Smith's usual attention and success.

Mr. Caird reports the Queen Charlotte to have been in perfect health since she left Spithead.

Sept. 22. Arrived the Thalia, Lord H. Powlett. Mr. Smith the surgeon reports her in good health. This ship left Sir A. Gardner on the 19th, all well.

Sept. 28. It appears that the small-pox have been brought on board the London by a child in its mother's arms. This method of communicating the small-pox infection to a ship's company, has been so frequent, that numbers of our seamen have on other occasions fallen victims to the disease. It certainly might be checked, by taking

care

---

the functions of life could be carried on to so great a length; one lobe of the lungs gone, and the other only half its proper size.

Juste, June.

(Signed)

T. KENNING, Surgeon."

care that no children of this defcription be
admitted.   Eleven cafes have already been fent
from the London to Haflar; as foon as the difeafe appeared : among thefe, is a young gentleman,
who had been inoculated when very young, and
though no eruption followed, the family apothecary affured the parents that the child was fecure.
This is a very delicate fubject in medical practice, and much concerns the happinefs of families:
we have met with many inftances of the kind, fo
that furgeons ought to be guarded, how they
decide; for in all doubtful cafes, it fhould be a
rule to inoculate the child again *.

   In another part of this work, under the article
SMALL-POX, I have propofed a general inoculation thoughout the Fleet ; the difeafe being of fo
ferious a nature, in fervice, from the number in
every fhip that never had it.

   On the 21ft of this month a contagious fever
made its appearance in the Glory, among the marines.   We could not trace it diftinctly to its
fource, but it was probably brought on board by
a draught that lately embarked.   The fymptoms
chiefly

-----

   * About this time the fmall-pox appeared in the Minotaur,
juft returned from the Weft Indies. The infection was communicated in the fame manner as in the London.

were chiefly cold shivering at first, sickness at stomach, sunk countenance, considerable stupor, weak pulse, and general debility.   Some cases at first were rather severe, but not fatal.

Mr. Carter the surgeon, being early aware of the nature of the fever, took the first opportunity of sending them on shore; by which precaution it did not show itself after the 29th.   On visiting the Glory, I found every thing so much to my satisfaction, that it was not deemed necessary to trouble the Lords Commissioners of Admiralty on the subject.   It was left to Sir George Home's choice, whether he would employ the usual fumigations, with the exception of not performing them when the people might be exposed on deck to very cold or moist weather: at the same time he was informed, that I had *no faith* whatever in this process.   The only means used to subdue the contagion, were, immediate separation of the infected, additional fires below, shutting the windward ports, and keeping the decks dry, &c.   It extended, in all, to thirty-four marines, and three seamen, who were probably hurt by hard drinking, having just received a large payment.

In the early part of the infection, there were cold easterly winds, with rain for two days; and it is not unlikely but the marines might be exposed as centinels.

M.   It

It is not much to be wondered at, that in the
fpace of a few days, a number of men fhould fall
down, when we confider that they fleep fo clofe
to one another, and expofed to the breath of each
other.   From the manner of flinging the hammocks, in the higheft part of the deck, they are
alfo the more furrounded by the leaft oxygented
part of the atmofphere :   the portion which
abounds moft with *azote*, is always found to occupy the higheft ftratum.

What do the advocates for fmoking and fumigating think of this fpeedy extinction of contagion ? Had the intelligent Captain of the Glory
ufed the *gaffes*, I fhould have fcarcely been able to
adduce it as a frefh fact, in fupport of my fide of
the queftion.   I apprehend the fame means in
fhips, equally well regulated, will always be fuccefsful : at leaft, it was found to be fo, in the
general infection after the firft of June.   Some
officers who tried fumigation with great earneftnefs
at that time, gave it up when they faw that J
confidered it as hurtful.

October 6. This day arrived at Spithead feven
fail of the line, the Niger and Migœra, under
Vice-Admiral Sir A. Gardner, Bart. from a cruize
off Ufhant.

The weather while this fquadron was at fea,
was moderate and fine; there were fome cold
                                              eafterly

eafterly winds, for a few days, and gales of wind commenced on coming to-port.

Thefe fhips are healthy after being feven weeks at fea.   The melaffes being now ferved to every fhip, the officers and furgeons fpeak of it with commendation.   It is ferved to the meffes, and much relifhed by the people: with it the oatmeal gruel is fweetened: the bifcuit by fome is toafted till it acquires the tafte of coffee, and when boiled and fweetened, makes an equally agreeable beverage: others prepare their peas in this manner. There has been very little fcurvy during this cruize, which muft have been prevented in a great meafure, by thefe additional delicacies in the diet.

Some catarrhs occured on board the Royal George; but Dr. Higgins found them yield readily to general remedies: The Royal George had feventeen venereal cafes cured at fea.

In the Namur, Mr. Seeds cured twelve bad fcorbutic ulcers, fince he left port.   Such a practice as this is vaft gain to the fervice.   He attributes his fuccefs to the citric acid, the molaffes, and the kind attention of Captain Whithed, in fending provifions and porter from his table, to whatever object the furgeon found them ufeful.

In the Defiance, Mr. Glegg cured twenty venereal cafes at fea.   The fhip's company were raw, yet only two cafes of fcurvy appeared. This

has

has been attributed to the method which Captain
Jones inculcates for the ufe of the melaffes.

The Niger has been ten weeks at fea: At leaving port, four relapfes in fever happened, among
men difcharged from Haflar. Mr. Burd does not
affign any caufe for thefe returns : no infection
fpread from them.

The Royal Sovereign, to which fhip Vice-Admiral Sir A. Gardner lately fhifted his flag and
fhip's company, fent only one man to the hofpital on coming to Spithead. Mr. Browne reports,
that *thirty-four* venereal cafes have been cured
during the cruize ; all of which had applied for
relief in the early ftage of the difeafe, and chiefly
imputed to the charge no longer operating as a
barrier.

In the Majeftic, catarrhal complaints were frequent, as reported by Mr. James Dunn. This
fhip returned but lately from the Weft Indies.
No fcurvy appeared ; the people have been long
together, and in other refpects were fufficiently feafoned for fervice.

Much valuable information comes from Mr.
Leggat of the Coloffus. This fhip had been lately commiffioned at Plymouth, and her complement compofed entirely of raw Irifh landmen, the
worft crew that ever came into a man of war ; a
great part of them had been difcharged from jails.
During the time of fitting, the febrile infection
made

made its appearance three or four times.   It only extended to a few cases at a time, and was kept under by sending 'the people to the hospital immediately.   One man became ill, and quickly all the members of his mess.   The ship in the mean time was well aired, and fires kept burning in different places.

Towards the conclusion of the cruize off Ushant, Mr. Leggat remarks, that the contagion appeared again.   I do not think this ship is perfectly secure, even now, when we consider of what kind of subjects her complement consists, and a season approaching that will give activity to the disease.   The scurvy has also made considerable progress in the Colossus, and must be expected to gain ground, should she go to sea immediately. She has, to prevent this, been ordered a double allowance of vegetables.

These helpless beings seem destined to feel the effect of every disease : during the cruize, near seventy cases of catarrh, unfit for duty, have been met with.

On board the Fame, the scurvy was scarcely known, till the arrival at Spithead ; when it appeared in a number of cases, but was quickly subdued by the fresh provisions and vegetables.

October 10. A number of ships having arrived from the West Indies, or expected at this time, and also from other foreign stations, I was in-

M 3   duced

duced to recommend to their Lordfhips, a fupply of vegetables for their ufe, fimilar to the Channel Fleet. This was readily complied with; and their Lordfhips have been pleafed to order, that I fhould exercife my difcretion in thefe, and all other occafions, in directing the quantity. The probability of hoftilities with Spain, when a number of new-commiffioned fhips would be filled with thefe people, rendered this meafure ftill more neceffary.

Nov. 10. I have mentioned above, my fufpicions that the Coloffus was free from contagion; and had recommended when on board, the careful feparation of the infected, with other ufual precautions. Fumigation is now difcarded from our fhips. To the beginning of this month, from her arrival, thirty cafes of typhus have been fent on fhore: fome with fymptoms of great malignancy. For the laft ten days none have appeared; and fhe may now be deemed fafe: fhe is juft ordered for foreign fervice.

Captain Grindall left the Coloffus at fea, from bad health. In a letter from Plymouth, he informs me, that two fever patients fent to the hofpital there, died; another died in the Sound, and feveral while at fea.: but the condition of this fhip, when he took the command of her, was particular. The Coloffus was the flag-fhip of Rear-Admiral Pole, and failed with the unfortunate

fortunate fquadron, under the command of Sir H. C. Chriftian, which was obliged to return from contrary winds and fevere weather, after being eight weeks at fea, in February.   The lower deck-ports could not be opened during this time ; and from the motion of the fhip, and water fhipped in the gales, the provifions and other articles had been fpread about the hold, to the quantity of forty or fifty ton, according to Captain Grindall's report, before the ballaft was expofed.   The confequence of this, was the production of an atmofphere in the hold and well, unufually foul : no doubt, owing to the fermentation of the provifions, and the decompofition of the moifture.   The officers and feamen employed in clearing the hold, were grievoufly afflicted with fwellings about the fubmaxilary glands, and violent opthalmias ; and were obliged to be relieved by others during this duty : fometimes there was a neceffity of ftanding faft, for fome days, that the noxious vapour might be expelled.   This account is interefting : but the fever certainly fprung from other caufes, and was brought on board by the people.

It is a little fingular, that the hold of the Glory, which was Admiral Chriftian's fhip, was extremely foul, and had not been cleared, when the fever made its appearance there.   I fhould have recommended the Glory to have been cleared,

had

had fhe not been under failing orders : but even in her, the fever originated from contagion im- ported ; otherwife, it could not have been fo completely overcome : the fhips returned from fea, report the Glory in perfeét health. But it is a part of our means for deftroying infeétion, to clear and dry the hold; and on that account, I have been more minute in this detail *.

Nov. 18th. This day the following fhips ar- rived at Spithead, under the command of Rear- Admiral Sir R. Curtis, Bart. viz. Formidable, Atlas, Cæfar, Mars, Triumph, Pompee, Orion, Irrefiftible; Stag and Proferpine frigates.

The Surgeons of the Fleet have received direc- tions from the Commiffioners of fick and wounded, to tranfmit to that Board a regular report of the medicine cheft, from furvey, with a view of fome additional articles being fupplied at the expence of government. I am forry that this bufinefs fhould be done, piece-meal : why not, at once, fupply every article, of both medicine and necef- faries, change the mode of paying furgeons, and modify their half-pay accordingly ?

Nov. 20th. The fhips juft returned from fea are in perfeét health, and have landed at the hof- pital, thirty people in different chronic complaints; yet

---

* Admirals Chriftian and Pole took with them, into other fhips, their own people ; fo that the Glory and Coloffus were filled up with raw men.

yet fome of them had been fourteen weeks from Spithead.   The furgeons, in their reports tranf-mitted to me, fpeak in terms of the higheft com-mendation, of the attention of officers, in every thing that relates to health.   During this cruize, very few cafes of fcurvy have appeared, as our preventive means feem to have attained nearly all the perfection of which they are capable.   We now hope, that the allowance of molaffes will be increafed, and made permanent.   The ftate of this fquadron affords a ftriking contraft to the condition of the fhips under Lord Bridport, when off Belleifle, in the fummer of 1795; and is the beft voucher for the utility of the meafures, which I propofed at that time.

On reviewing the occurrences of this Fleet, it muft be a pleafing reflection to Government, to be convinced that the late encouragement given to furgeons, has been extended to men, who merit every thing for their abilities, attention, and fide-lity: the beft teftimony of thefe qualifications, is the health of their fhips, and an empty hofpital.

The naval character has been long renowned for the *heroic virtues :* the tranfactions of our fhips will demonftrate, that it has attained the higheft ftation in thofe that are *amiable*.   Great Britain, thus defended by her Fleet, and the health of her

feamen preferved by the benevolence of her coun-
cils, will continue to wield the trident of the ocean,
and engrofs the commerce of the world. War
itfelf affumes a new afpect :. thofe difeafes formerly the fcourge of a fea-life, are prevented or
overcome; even Contagion, that has fo often
fpread terror and difmay, has, by a fyftem of duty,
made familiar to officers, become no longer terri-
ble! With thefe reflections, I muft clofe my detail
for the prefent.

MEDICINA

CONTAGION.

THIS subject is one of the first magnitude, that falls to the discussion of a Physician. By it, the operations of fleets and armies are overthrown on a sudden, and the best concerted plans rendered abortive. Every fact that relates to it, becomes valuable, insomuch as it assists our vigilance and discernment to detect it when concealed, or suppress it when apparent. The preceding part of this work, though comprehending only the naval transactions of a short space, has afforded an ample proof of what serious consequences may be expected from the introduction of infection into a ship: when passed over with indifference, we have seen it run the hazard of disabling a large part of our fleet for a length of time. But to treat it with levity and neglect, is criminal in the highest degree; the lives of human beings are the victims of this obstinate folly. Ignorance itself is unpardonable on this subject;

§   for

for every phyſician and ſurgeon employed in the naval or military department, ought to make themſelves acquainted with all that has been written on CONTAGION, and keep a regiſter of their own obſervations, in order to familiarize the knowledge of it to others.

It is of great practical importance to trace in-fection to its ſource, the means that have conveyed it, and the cauſes that have given it activity. Though on an unpleaſant buſineſs, a mind ac-cuſtomed to enquire after truth, will ſometimes find no ſmall degree of curioſity and amuſement. Some of my own adventures have now and then been romantic, if not hazardous, for the purpoſe of putting queſtions to a patient, or to clear a doubt. It is only from being minute and exact, that we can arrive at ſufficient information; and whoever may attach himſelf to the inveſtigation, will conſtantly meet with ſomething to encourage his perſeverance, and enliven his reſearches.

The field for obſervation, which has been em-ployed by my ſtudies, has not been confined to a naval life, and acquaintance with the hoſpitals only; but an extenſive practice in a country ſitu-ation, and a few remarks made in a large town.

What I have to offer, may be conſidered as an abridgement of general facts, rather than a me-thodical arrangement of them.

I would

I would define Contagion, to be a something propagated from diseased bodies, or from substances that have been in contact with them, producing a similar disease in other persons.

The latter part of this definition includes what have been called *fomites*, whether we consider them as wearing apparel, bedding, or other articles that have been imbued with human effluvia, of persons labouring under infectious diseases. I have made use of the word *something*, as contagion becomes only familiar to our senses, as a something impregnating, or conveyed in the exalations which proceed from bodies under actual diseases, or what have been in contact with them *.

There are causes and circumstances which favour the propagation of contagion in the diseased subject, and increase its virulence when generated.

There are also causes and circumstances which favour its reception in the healthy body.

That there is a state of body, in a contagious disease, where certain causes and circumstances contribute to increase the quantity, as well as the virulence of contagious matter, appears to me very obvious. A more aggravated degree of malignity, as it is called, will generate a greater quantity of infection, and as it may be confined in a larger or smaller space, it will be less or more

* I wish to be understood as speaking of Typhus Contagion alone, unless otherwise expressed.

noxious

noxious. The expreffion, Malignity, often ufed in fpeaking of this fubject, ought to be explained. I would call a fever malignant, when with the fymptoms of debility, there is a cadaverous fmell arifing from the body, an unufual fœtor of the breath, ftools, and other excretions, the tongue black and parched, the eye dufky or yellow, the countenance bloated and dejected, and the fkin fallow. In approaching a fick bed of this kind, a perfon not much accuftomed to fuch vifits, will be very liable to receive the infection; and the unpleafant fmell will be much fooner perceived, than by the phyfician, or other attendants.

Variolous Contagion, we can well fuppofe to be more rapidly fpread from a patient with a large number of puftules, than from one with only a few; fo alfo the confluent kind will be more hazardous than the diftinct, and by whatever means the one is converted into the other. It may be objected to this, that pus taken from the confluent kind, will, by inoculation, produce the diftinct fort; fo may it likewife by the natural method: but this does not do away the argument, that the one fpecies generates more contageous matter than the other. The virulence of variolous infection, like that of typhus, will be increafed, in proportion to the fpace in which it is diffufed, whether in the atmofphere, or in matter taken from a puftule. By diluting the firft
with

with a large proportion of air, it gradually lofes its power of communicating difeafe. The other becomes inert by exficcation, to a certain degree; but when foftened by water at blood heat, fo as to be applied beneath the cuticle, it regains its activity, unlefs too much diluted, when it lofes it altogether. This virulence of contagion, is, therefore, nothing more, than the exhalations of the fick, diffolved in a fmaller portion of atmofphere.

We draw the conclufion, that a malignant typhus is more apt to generate contagion, becaufe we fee that flight cafes of the fame difeafe, and with mild fymptoms, do not extend to others, although no means of prevention have been ufed; and in fituations too, where there were a ftrong pre-difpofition to affift its action.

There is alfo a period of the difeafe itfelf, that is incapable of generating infection fufficient to produce the fever in others. We are affured of this fact, from a timely feparation having prevented the further progrefs; and by this means alone, I apprehend, we eradicate contagion in fhips or any where elfe. We cannot draw any line of certainty, at what time it may ceafe to be fafe in permitting people to affociate with a patient; it will depend moft on the nature of the fymptoms, whether they are mild or malignant.

In the fmall pox, the difeafe feems incapable of infecting another perfon, before the fecond or third day of the eruption. This has enabled us to

remove

remove patients into the hofpital and hofpital-
fhip, after the difeafe was afcertained, fo as to
fecure others from being affected. We have feen
this in the Gibraltar, Valiant, Queen Charlotte,
&c. In meafles, however, this does not feem to
hold good. The difeafe may be propagated at
the moft early ftage of eruption ; and if I was to
be allowed a conjecture on the fubject, I would
fay, that the contagion is the offspring of the
catarrh which accompanies the meafles.

Subftances imbued with the exhalations from
infected bodies, if not expofed to the air, have
their powers of communicating difeafe increafed ;
or in other words, the infection from *fomites*, is
faid to become more virulent than it was when
firft feparated from the body.

I am of opinion with others, that the exhalations or excretions of the fick, are the vehicles
of contagion. It is thefe which impregnate the
atmofphere with noxious matter : they affect in
like manner, bed-cloaths or apparel, and every
thing that can imbibe them, when in contact with
the difeafed body. When bed-cloaths, or body-
linen, but particularly filk or woollen cloth *, have
been expofed to thefe exhalations, and then heaped
together for a length of time, the noxious
effluvia are, as it were, multiplied, and will more
certainly infect others, than they did at firft.

The

* Being animal matter, and more eafily liable to decompofition.

The bales of goods which brought the plague to Marfeilles, and affected the people that opened them fo fuddenly, had their virulence increafed by not being duly ventilated. When the jail fever contagion was brought into court by the prifoners at the Oxford affizes, and more lately at the Old Bailey, the fever was propagated from the cloathing of the prifoners : no doubt, from being confined in impure, ill-aired cells, this infection became more virulent. The highly concentrated ftate of the contagion in the bales of goods, could only have been brought to that ftate of virulence, from the clofenefs of the package; it cannot be fuppofed that any human beings could have put them together otherwife. The nurfes of hofpitals know well, as Dr. Lind tells us, that there is moft danger of catching a fever, when they pile heaps of bed-cloaths or body-linen together, for a few days, before it is carried to the wafh-houfe. The wafherwomen at Haflar have alfo told me the fame thing. They know when a dangerous fever is in the hofpital, from the bad fmell of the cloathes; this makes them air them abroad, till the fmell is gone, and then they can wafh them with fafety. But if it happened, from the hurry, that this could not be done, or if it was neglected by defign, many of them have been feized with the ficknefs. The porters and people employed in cleaning and fumigating the blankets and beds at Haflar, are well

N   acquainted

acquainted with this fact, and they meafure the
danger by the badnefs of the fmell. This ought
to inftruct every body to ftand to windward of
thefe infected fubftances, when they are opened;
as the current of air would then carry it the other
way. In one of the courts of juftice, the people
who ftood between the prifoners and a window into
which the wind blew, efcaped the infection, while
thofe on the other fide were fufferers.

In the fummer of 1793, while the Oreftes brig,
commanded by Lord Auguftus Fitzroy, lay at
Plymouth, fhe was anchored very near and to
leeward of an army tranfport, which had on board
a very malignant fever among the foldiers. While
the foldiers were moved on deck, to go on fhore
to the hofpital, the crew of the Oreftes, from
curiofity, walked on deck, to look at them. Such
was the concentrated ftate of the contagion,
among the cloathing and bedding of thefe troops,
on coming from below, that eighteen people belonging to the brig, were quickly feized with the
fame fever; the infection of which had been conveyed by the current of wind. It did not, however, extend much farther in the Oreftes, from the
attention of her Commander. But this ought to
be a caution for fhips to keep clear of thofe that
have fevers on board, as a virulent contagion may
be conveyed for a confiderable diftance.

x   Dr.

Dr. Lind is inclined to think, that washing the bed-linen in hot water, even when first shifted, is attended with much risk; and that the noxious matter may be volatilized by the heat of the water, and affect the woman. For this reason he has recourse to his favourite process, *fumigation*, to insure the washerwoman. The heat of his fumigating furnace, would no doubt dry the linen, and exhale any moisture: but our practice in the Charon, was to plunge every thing, as it came from the bed, into a tub of hot water kept ready on purpose. The linen was washed and dried immediately after. We had in that hospital many malignant cases of typhus, and some deaths, yet no infection was ever spread there.

Let us now inquire how infectious matter becomes more virulent, and liable to spread disease, when confined, as in beds, cloathing, or bale goods, &c. denominated *fomites*.

This fact being well known, did not fail to excite the attention of physicians to solve the mystery. The most plausible reason that could be assigned, was the generation of animalcula; the cotton or woollen cloathing was said to serve as a nest for the corpuscles to multiply; and thus the contagion was thought to increase sevenfold. We shall afterwards see that this notion entirely regulated the different methods that have been used to destroy contagion, and check the fevers

N 2

which

which were fpread by it. This mode of account-
ing for the fact, was certainly a bold effort of the
imagination; but I do not think it can be called
any thing elfe; for thefe little animals have never
been feen even with a microfcope; and I cannot
help concluding, that like fome other animals
which we have heard of, they are fabulous. The
fubftances we are now fpeaking of, being more or
lefs tainted by the excretions of the fick, it will
facilitate our inquiry to know the nature of thefe
excretions. Now the fœtor of the breath, perfpirable matter, &c. evidently demonftrate that
they differ from the healthy ftate. The fmell, to
our fenfes, comes very near to what is called *ful-
phurated-hydrogenous gas*. Some of the fluids within
the body would feem to be, in fome degree, in a
ftate of actual decompofition; unlefs we can fuppofe the mucous glands of the lungs fecreting a
fluid, that taints the expired air in this manner.
The decompofition of the fat, which fometimes
difappears very fuddenly in fevers, may give fome
ground for the fuppofition, that a large proportion
of thefe exhalations are compofed of hydrogenous
gas. But whether we can go thus far or not, what
is feparated from the body, it is plain, is more difpofed to decompofition, than when the body is in
health. Now this procefs will ftill go on, whether
expofed to the atmofphere or not, with this difference, that, by expofing fubftances which have
imbibed

imbibed the exhalations of the difeafed to a free air, the noxious gaffes will be diffipated as quickly as they are evolved : while on the other hand, by laying the cloathes in a heap, packing them firm in a cheft, or making up cloth into bales, the gaffes are concentrated into a fmall fpace ; *qua data porta ruunt*, and woe to the man who firft infpires them !

This appears to me, to be a better explanation of this bufinefs, that fubftances tainted by the bodies of the fick, are more liable to communicate difeafe than the bodies themfelves.   Now this does not hold out an idea, that the powers of contagion are multiplied, as by generation; for that would be to fay, that thefe gaffes are *themfelves*, what we call the matter of infection.   I would only go fo far as to affert, that they are the vehicles of it, till more certain experience fhall determine farther.

It has been long fuppofed that contagion may be engendered in particular fituations ; and typhus fever is faid to be often the confequence of this. When a number of people are confined in a fmall fpace, ill-ventilated, damp, and dark, to which may be added, neglect of cleanlinefs, both in perfon and cloathes, with low diet, and depreffion of fpirits, a jail fever often follows.   When the human body is expofed to air not fufficiently oxygenated, the faculties of mind, as well as the powers of the body, lofe their proper vigour.   The body, in this

N 3   cafe,

cafe, affifts in fowing the feeds of difeafe for its own deftruction. The deficiency of the due ftimulus, leaves the body in a ftate of debility and torpor: the blood lofes its florid colour, and the mufcular fibre its power of contraction. By thefe means the fecretory organs no longer feparate a healthy excretion, and the air is further injured by a fœtid breath and perfpiration. If intemperance in the ufe of fpirituous liquors, is a part of the habit of living, it joins its effect to that of other hurtful caufes; and hard labour, and a watery aliment, bring the fubject to the degree of typhus debility.

Such is the hiftory of human mifery in many of the great towns of England. And now that I am upon the fubject, it would be a fin to pafs without unfolding it a little more. During the fhort time that I attended the difpenfary of Newcaftle, juft at the beginning of the war, I was fent for to a poor man, in a miferable and low part of the town, called Sandgate. He was ill with what is called a fpotted fever. He lay in a large garret room, on a bed without curtains, round which were ftanding fix children, the oldeft not more than nine years old. I afked the ufual queftions concerning his cafe, and whether he had any perfon to affift him? He anfwered in a tone of voice, that befpoke the deepeft affliction, that his wife was gone out, for the firft time, fince fhe recovered from the fever,

(the

(the children had been ill firſt) to ſell the laſt gear
they had to ſpare, to *buy meal for the children!*
He worked on the quay, and his wages were
one ſhilling and two-pence per diem, for the year.
This man walked to my lodgings in a fortnight,
the firſt day he returned to work; for he ſaid that
he could not afford to lay by any longer.

When I practiſed as a ſurgeon and apothecary,
after my return from Africa, at the end of the late
war, in a ſmall town in Northumberland, with an extenſive country buſineſs, ſome ſcenes ſimilar to what
have been juſt related, came under my view. Two
ſervants, of two opulent farmers, applied to me for
relief. The firſt had ſeven children, who took the
fever, one by one, till the whole became ſick. His
wages were one ſhilling per diem. His maſter was a
rich man, and well knew the wretched condition of
this family. He thought himſelf charitable, by allowing them to pull turnips from his field, *for food:*
thus they fared as well as his oxen and ſheep!

The other ſervant was a ſhepherd; but his herd-
ing, as the ſaying is, was a poor one. The firſt and
ſecond of ſix children, were able to work a little, till
they got a fever in a ſevere winter, and down they
fell, one after another; the father and mother at laſt.
They wanted to ſell the cow, which they thought
would buy neceſſaries while they were ill, and pay
the doctor: and by this time warm weather would
come, and the oldeſt would be able to work. To

N 4 prevent

prevent this, a fmall fubfcription was obtained
from fome charitable ladies in the neighbourhood;
by which means the comforts of wine and diet
came within their reach. In the mean time, the
mafter heard that a fum of money had been raifed
for the relief of the family. His heart opened—
a rotten fheep was found dead in a furrow; he defired the fhepherd to ufe it for his children, but to
take care *to return the fkin!* * Hear this relation,
ye legiflators, and fix the *minimum* of labour!—
Thefe cafes are a few of many fcenes of difeafe and
poverty, which have fallen to my attendance.

The fituation of the African negroes, confined
during the middle paffage, in the flave-rooms of
a guineaman, has been mentioned by Dr. Lind.
The confinement of fo many wretched creatures,
in a fmall fpace, defervedly attracted the animadverfion of a phyfician inveftigating the fources and
progrefs of Contagion. But contagious fevers we
find are not their difeafes. We can well believe,
that if the negroes were cloathed, that filth and un-
cleannefs might generate infection; the exceffive
quantity of perfpirable matter emitted from the
furface, in a high degree of heat, would foon accumulate, and adhering to linen or woollen cloth,
might at laft propagate the forms of difeafe. But
this matter being daily wafhed from their fkins,
and the rooms kept clean, nothing offenfive, or of

* What a contraft to the benevolence of our officers!

animal

animal origin, is allowed to undergo the final decompofition, which it would do, in nafty and un-
ventilated cloathing.   Thus alfo the poor inhabitants of warm countries are free from the difeafes
of thofe in colder regions.   Indeed, we much doubt,
if a genuine typhus has been ever feen within the
tropics: it has not been often feen.   A Liverpool
fhip, called The Hankey, by being crowded with
people, at Bulam, on the coaft of Africa, was infected by a contagious fever, of which many died.
Dr. Chifholm dates the origin of a fever at Grenada, from this fhip.   But his defcription does
not correfpond with our jail fever.

Does marfh effluvium differ from contagion, or
in what properties do they refemble one another?

It has been argued by phyficians of the firft authority, that intermittent fevers arife from the
miafmata of marfhes, and continued fevers, from
human effluvia.   This is no doubt a very general
rule, but it admits of many exceptions.   I have
feen a family, where the father was labouring under
an obftinate tertian, while the mother and fome
of the children were ill, in bed, with a typhus.   Yet
the remote caufes, as near as we could trace them,
were the fame in all.   They proceeded, in the
month of February, from a cold damp houfe, and
deficiency of diet.   Nay, I have conftantly remarked in thofe fhips, where contagion prevailed,
that many cafes of regular intermittents, and remittents, occafionally appeared.   When they increafed

creafed in proportion, to the number of the continued type, and the latter becoming milder in
its attack, I confider it as an infallible fign that
the power of the infection was on the decline, and
would be fpeedily fubdued.

I fhall now give fome inftances, where continued and intermittent forms, appeared promif-
cuoufly during the progrefs of one contagion.
On the 12th of February 1793, I joined the Vengeance of 74 guns, at Spithead, fitting to receive
the broad pendant of Captain Charles Thompfon, for the Leeward Ifland ftation : and alfo to
carry out five gentlemen, deputies from the French
Iflands to the Britifh Miniftry.

The Vengeance had been a guardfhip at Chatham, and was ready to proceed to Portfmouth,
in the beginning of January, fully manned. At
this time fhe received four hundred men from
different yeffels, to carry to the fleet : of this
number, came two hundred from the Nemefis
frigate, at the Nore ; and among them, two men
ill of the jail fever, one of whom foon died. The
paffage to Spithead was long, and the weather
rainy and boifterous. During this time, the infection, from the two fever patients, had made a
great progrefs ; for here was every thing that
could render it active, a crowded fhip, new-raifed
men, badly cloathed, fleeping on deck, a cold
feafon, &c. Sixty cafes were fent to Haflar on
the

the arrival at Portfmouth ; fome had been cured on board ; and when I joined the fhip, there were eleven people confined to their hammocks with this fever.

The difeafe was hitherto almoft confined to the new-raifed men, who were Irifh landmen, dirty and ragged to a proverb, the very dregs of jails in the metropolis. Men of this defcription, fuch as are often to be met with in fhips, are the victims of all general ficknefs. They poffefs neither the courage of an Englifhman or Scotf-man under difeafe. They fink from defpondency when really ill ; while at other times, with a fer-vile low cunning, they are conftantly peftering us with trifling or pretended complaints.

The ufual precautions of purifying the fhip were immediately put in execution, and addi-tional cloathing was ordered to thofe in want of it ; at the fame time care was taken to enforce perfonal cleanlinefs where it was neceffary. Every man, with the firft fymptoms of indifpofition, was carefully watched, and fent on fhore.

About the beginning of March, the infection feemed to be on the decline ; which appeared from the attacks being lefs fevere : fome cafes of the remittent and intermittent type, were now obferved. Trufting to a mild feafon approaching, and the voyage we were bound, to warm latitudes, I confidered the fhip's company in condition for fea. The Captain tranfmitted this opinion officially

to

to the Admiralty ; and we were ordered to depart, without delay.

On Sunday March the 10th, we failed from Spithead with a convoy. The fick lift confifted of feventeen venereal patients, and feven in fever, fome of them regular tertians ; and to thefe may be added a few with bad ulcers, who were kept on board, being valuable feamen *.

The weather was cold ; the wind at N. E. and the thermometer at noon, 44° of Farenheit.

On the thirteenth we arrived at the Cove of Cork, with two frefh cafes of fever. During our ftay here, the lift continued to fluctuate ; fome were every morning difcharged to duty, and others added in the courfe of the day ; in both the continued and intermittent forms.

March the 23d, failed from Cork, with orders to join a fquadron under Rear-Admiral Gardner, deftined for the Weft Indies ; it having been reported that the French had difpatched fix fhips of the line, to thefe.feas. From this time to the 31ft, we cruized off Scilly, without meeting Admiral Gardner ; when the Captain thought proper to bear up for Spithead. The wind for the firft

---

* It was fome time before a fever of this kind was fufpected on board the Nemefis, after fhe had fent her fupernumeraries to the Vengeance. She was now come to Spithead, and obliged to go into harbour, having fent nearly her whole complement to the hofpital.

fix days was tempeftous at N. E. and very cold,
but fhifted on the 31ft to S. W.   The attacks
of fever, fince leaving Cork, had been more numerous than on the preceding week.   They were
both of the continued and intermittent types; in
all feventeen.

On Tuefday the 2d of April, arrived at Spithead.

While in port, the lift varied occafionally, but
cafes of fever ftill appeared.   On the 16th of
April failed again.   On the 28th anchored in
the Cove of Cork ; and failed for the Weft Indies on the 2d of May, with a large convoy.
Thermometer 49°. Sick lift twenty-three ; eleven
of thefe, fevers and agues.

May 6. Some fever patients returned to duty ;
two others added to the lift. Therm. 59°.

May. 10. Lat. 45,21°. Therm. 60°. Fevers
doing well.

May 15. Lat. 32,36°. Therm. 62. Only two
fevers added to the lift fince the 10th. Both cafes
with regular remiffions.

May 17. At Madeira. Therm. 66°. No fevers
confined to bed.

The ulcers, from great attention to dreffing and
diet, getting faft well.

May 25. Lat. 25,25. Therm. 69. Only three
very flight attacks of fever fince the 17th.

Sick

SICK LIST.

Convalefcent - 6
Fever - - - 3
Venereals - - 3
Ulcers, &c. - 9

21

May 27. Lat. 23. 34. Therm. 74. Croffed the tropic in the afternoon, when the failors performed the ufual ceremonies. Neptune in his car was drawn round the deck by fix tritons.

May 31. Lat. 20,14. Therm. 72. Sick lift thirteen.

June 4. Lat. 15. Therm. 76. This day a man died, who had the fever fix weeks ago : but feemed to fink under mental affliction.
No Fevers.

June 11. Anchored in Carlifle Bay, Barbadoes. Therm. 82. Sick lift ten. But neither a feverifh patient, or venereal, or ulcer, of the number.

Had we been fortunate enough to proceed on our voyage when we firft 'left Spithead, the contagious fever had probably difappeared long before it did. It was curious to obferve how quickly it declined as we approached the warm latitudes. The treatment of this fever, and particularly the agues, will be given under their refpective articles.
This

This ship, the Vengeance, returned home in October, in perfect health, except what she suffered, in a paffage of nine weeks, from scurvy.

Some time in the months of October or November 1793, the army under the command of Lord Moira, embarked; for the purpose, as it was fuppofed, of making a defcent on the coaft of France, to affift the loyalifts. Part of thefe troops had been originally intended for an expedition against the Weft India Iflands, under Sir Charles Grey. But the deftination being changed, and the paffage expected to be fhort, being only the fmall diftance from the Ifle of Wight to the coaft of France; it was thought that two tranfports could fafely contain what were in three. Health on this occafion, and feafon of the year, feems to have been little attended to. This army proceeded to Guernfey, not being able to land on the French coaft: fome of the tranfports were driven from their anchors in gales of wind ; and the whole returned to England in the end of December, very fickly. The weather was boifterous and wet all the time ; which, with the crowded ftate of the veffels, foon fpread fevers and dyfenteries among the foldiers. The accommodations at the Ifle of Wight, being unequal to the number of fick, a great part of the worft cafes were fent to the Royal Hofpital at Haflar, where I

was

was now appointed Phyfician.   Whether the infection had been carried on board at the time of
embarkation, I was not able to learn : the foldiers
themfelves, imputed their illnefs to the crowded
dirty tranfports, and the confinement during bad
weather.   Many of them had not taken either
medicine or food that was fit for their fituation ;
for fome of the fhips had neither furgeon or mate.
There were a larger number of bad cafes in typhus, ague, and dyfentery, than come ufually to
a naval hofpital at one time.   From the fame
tranfport, and in the fame regiment, were brought
people ill of the three difeafes juft mentioned :
and the general remote caufes, from every en-
quiry that I was able to make, appeared to be
the fame in all.   A few of the agues long refifted
the powers of medicine.

We have now feen in the Vengeance, the contagion of typhus, brought on board by two men,
extend throughout a fhip's company : and in the
fpace of four months, in different fubjects produce
continued remittent and intermittent forms of
fever.   In Lord Moira's tranfports, fomething of
the fame kind occurred, with the addition of
dyfentery.   To thefe I might add other facts in
confirmation ; from almoft every fhip which I
have attended under a general contagion.   Vide
Mr. M'Callum's Report in the General Abftract,
for May 1794.

Marfh

Marſh miaſma ariſes chiefly from ſtagnant water; or its decompoſition, occaſioned by animal or vegetable putrefaction: for water ſuffers no alteration, unleſs ſubſtances are thrown into it, which facilitate the ſeparation of its principles. The marſh effluvium, like contagion, acts more powerfully, in proportion to the quantity of at- moſphere in which it is held. Hence its influence is directed by the current of wind; and the nearer you approach the ſpot where it is generated, the more certainly will the body be affected. Thus, houſes near ſtagnant pools, by the opening of windows, have had their inhabitants ſeized with agues; and camps, to leeward of marſhes, have ſuffered in a ſimilar manner; or eſcaped by mov- ing to another direction.

Hydrogenous gas being copiouſly evolved du- ring the decompoſition of water, would make us ſuppoſe that it is a principal ingredient in this noxious effluvium. If vegetable matter has aſſiſt- ed the proceſs, of courſe a portion of carbonic acid gas will be added: but there are ſome vege- tables which contain azote, and when they are immerſed in ſtanding water, it is well known that the ſmell becomes uncommonly offenſive. Such are the water of putrid cabbages, and the ponds where flax has been ſteeped to effect the ſepara- tion of its rind. I have frequently known this

O water

water prove deleterious to cattle. Animal fubftances muft therefore be more hurtful than vegetables. Salt water, in contact with vegetable
matter, when confined in pools, becomes fooner
offenfive than frefh water; but it would be difficult to follow thefe ingredients through all the
different forms of combination. Thefe facts,
however, ftrengthen the fuppofition, that hydrogenous gas is the chief agent in marfh effluvium.
I would call it nothing more than fuppofition,
becaufe the ultimate decifion is not to be obtained; it becomes too fubtile for our fenfes.

What has been called the infenfible perfpiration, which arifes from the furface, and the lungs,
we have a right to believe carries with it in folution a portion of the variolous matter, which
charges the atmofphere with the contagion of
fmall-pox; even in fuch quantities as to impregnate the cloathing of attendants and vifitors; by
which means it has been frequently carried to
families and villages many miles diftant from its
fource. Different difeafes generate their fpecific
contagions by different methods; as elephantiafis,
lepra, lues venerea, pfora, &c. Even the rabies
canina has its organ of poifon, for the faliva becomes the vehicle of infection.

The petechiæ, vibices, or miliary eruptions, and
apthæ, which are common in malignant fever,
form no part of the character: fometimes a few
                                                    of

of them appear, and at other times nothing of the kind can be obferved. We know of nothing that can propagate this fever, but exhalations from the body. But a patient in typhus was fent from the Venerable to the hofpital fhip, with a fœtor about him, that exceeded any thing of the kind which ever came within my knowledge. After being wafhed and fhifted, it ftill continued; and was perceived at a confiderable diftance. He died in a few days; yet nobody was infected from him, either in his own fhip, or our hofpital. There was probably fome peculiarity of conftitution here.

We fhall now take notice of thofe caufes and circumftances which favour the reception of contagion in the healthy body. This includes both external agents, and what has been called predifpofition.

The moft prominent features in the character of the typhus contagion, are its generally appearing in a cold climate or feafon, and difappearing as fummer advances. This has been long obferved; but particularly mentioned by Dr. Blane. There may be many exceptions to the general rule; but the fact feems fully eftablifhed. My obfervations have all been to this effect; that the infection of typhus is moft apt to prevail in cold weather; and if it fhould appear in fummer, it is more eafily fubdued.

Among

Among the poor, in large towns and cities, who live in low, dark, and damp houses, it begins in the months of October and November; and disappears in April and May. Among the indigent labourers in country situations, where I have met with it, from the same causes, its beginning and termination vary little from this form. Even in ships, it proceeds, more or less, in that manner. We have constantly remarked, that the number taken ill in a given time, depended very much on the state of the weather. A few rainy days, in succession, never failed to increase the sick list. The people sent to Haslar from the London, complained of the weather to which they were exposed in returning the ship's stores: no particular disease was known in the Gibraltar, till she encountered strong gales of wind, with rain, in attempting to join the Fleet. The fever patients in the Vengeance were uniformly more numerous, exactly in correspondence with the weather being wet and stormy: we also observed this in the Valiant. The army under Lord Moira, when embarked at Southampton and the Isle of Wight, were thought fit for actual service: but the weather no sooner became wet and boisterous, than the sickness, typhus, and dysentery, spread in proportion.

When I first visited the Portugueze squadron at Spithead, there was not a man confined to
bed;

bed; but I faw in each fhip a few fickly-looking men walking about, evidently with flight fymptoms of typhus. It was now fummer, and the weather uncommonly warm in England. It did not require much foreknowledge to predict, that thefe people were not equal to fea-duty in our Channel. We left St. Helen's in the beginning of September; and, on the 1ft of October, there were upwards of four hundred cafes of fever among them.

Now, as cold weather, and a winter feafon, favour the action of typhus infection; we know that warm weather, and a fummer feafon, affift in its extinction.

The fever which was fpread from the French prifoners to our feamen, after the battle on the 1ft of June, appeared in the firft attacks about the 8th and 10th of the fame month. There were, perhaps, never any inftance of contagion becoming fo general in a fhort time. Moft of the French prizes had a number ill of the fever, particularly the Sans Pareil, Northumberland, and Vengeur, which fhip funk a few hours after the action; from the laft the Alfred was infected. More than two thirds of our Fleet were by thefe means expofed to the contagion. The heat was greater at this time than ever I remember to have known it: this alarmed many for the fafety of the Fleet. To the extreme heat, however, I at-

tributed,

tributed, in a great meafure, the fpeedy extinction of the fever: under other circumftances of feafon and temperature, I think our fituation might have been very different.

In the cruize with Lord Bridport, off Belleifle, in the beginning of July, a fever of this kind appeared on board the Orion; which was firft known by eleven men being fent to the hofpital fhip. I traced the contagion to men, who had come from the Royal William receiving-fhip, a few days before the Orion failed, as mentioned in the general abftract. The weather was very cold in the firft part of our time at fea; and I think that foftered the infection: but the fummer heat was now confiderable. Every man with the leaft fymptom being moved on board the Charon, not a fingle cafe appeared afterwards: three of thefe died. I could not help believing, that the warm feafon had a confiderable fhare in aiding our endeavours on this occafion; becaufe the Orion had a new fhip's company, and not quite brought to due order and difcipline.

The fever which appeared in the Niger frigate, affords another example of warm weather favouring the expulfion of fever. Captain Foote, who watched over the health of his people with the concern of a father, thought of nothing but being obliged to move every article out of his fhip, to eradicate the difeafe. There occurred a few warm days about this time, which did not allay his

fears,

fears. Knowing that the Niger was under the firſt order and diſcipline, I felt no alarm; but deſired them to continue the ſeparation of the diſeaſed, as ſoon as a freſh attack was known. This, with other attentions on the ſide of the officers, and Mr. Burd, the ſurgeon, cleared her in a week, after ſending forty-four men on ſhore.

The voyage of the Vengeance to the ſouthward, may be juſtly cited as one of the ſtrongeſt proofs we can draw; and by referring to it, I ſhall proceed to ſomething elſe.

We ſhall now ſpeak of the ſtate of the body, that ſeems to favour the reception of contagion.

Authors ſeem to have agreed, that there is a period of life, in which fevers are more frequent than at another. After the age of forty-five, the chance is in favour of the conſtitution eſcaping diſeaſes of this claſs. The reign of health and ſtrength, therefore, appears to be endowed with ſomething that renders the body ſo eaſily ſuſceptible of contagion. This is that ſenſibility of the nervous ſyſtem, which prevails in youth, and is gradually obliterated as we approach to old age. It is connected with certain external ſigns of character: ſuch as a blooming florid complexion; ruddineſs of the whole ſkin; arterial plethora, and muſcular ſtrength.

It is, however, to be obſerved, that it is in particular conditions of body, in the youthful conſti-

tution,

tution, that efpecially renders it liable to be affected by contagion, more at one time than at
another.   This is that ftate of debility, that fucceeds to all præternatural excitement; fuch as
fatigue after labour; the languor which follows
debauch, as hard drinking and exceffive venery;
cold after being over heated; approaching the
fick-bed with an empty ftomach, fear of being
infected, &c.   We can only reafon on thefe predifpofing caufes, as leaving the body in a ftate to
be acted upon by the contagion.   But the condition of people after intoxication, is a more general remark than any other; and when it is combined with other debilitating pleafures, is ftill
worfe.   Many of Dr. Cullen's pupils will remember his facetious application of this fact.   The
Doctor faid, when he was called to vifit a ftudent confined with a typhus fever, he knew very
well how it was got. " My young gentleman," faid
he, " after getting drunk, and paffing the night
in houfes of *a certain defcription*, comes the next
day, debilitated with debauch, into the clinical
ward, where he is expofed to the contagion."—
The laft feafon I had the honour to attend my
venerable Preceptor, at the very time of the lec-
ture he was giving on this fubject, three gentlemen of my own acquaintance lay ill of the fever.
After fome irregularities of the preceding night,
they all received the contagion at the bed-fide of
a hand-

a handfome fervant girl, that lay dangeroufly ill in the infirmary; whofe youth and beauty made them pay her particular attention at their vifits.—— Dr. Cullen would add, " that he generally found new married people among the number infected, when fevers prevailed in a neighbourhood."

While the Fleet were infected in the fummer of 1794, a part of the prize money was paid about the beginning of July. It was particularly hurtful to the feamen, and tended in a great degree to multiply the cafes of fever. *

Thefe conditions of body may alfo be applied to the action of marfh miafma. Hence fleeping on the ground in a ftate of intoxication, has been fatal to many in low damp fituations; but the thoughtlefs failor and foldier are the moft frequent

* " At laft report, I had the fatisfaction to remark, that the fever, which fo long harraffed our people, had very nearly, if not entirely, difappeared; and, previoufly to the late battle, we had become tolerably healthy. Since that period, however, a fever equally virulent, or even more fo, has appeared in our fhip, and ftill continues to rage. There can be no doubt, that we owe this frefh importation to the French prifoners we received after the action of the 1ft of June; as I obferved feveral of them affected with it, a day or two after they came on board; whom I have no doubt had been ill fome days before I faw them. At this time our people were apparently free from it; and I mentioned to Captain Pringle my fears of the confequence. Notwithftanding every thing was done to prevent, as much as poffible, all communication, I am forry

quent victims of this indiscretion; hence the fatal diseases which destroy them in the East and West Indies.

The withdrawing of any accustomed stimulus from the body, also favours the action of contagion, and the effluvium of marshes. Thus, a sailor thinks he will escape a fever as long as his tobacco lasts, and dreads the end of it. An empty stomach is hurtful when exposed to either disease. Bleeding and purging are hurtful; they act by leaving the body rather in an absorbing than perspiring state.

The application of cold to the body; whether from the surrounding atmosphere, or from cold water, or wet cloathing, and whatever gives the sensation of chillness to the surface, favour, in an especial

•

forry to observe, my prediction has been too much confirmed. It has unquestionably been much aggravated by the intemperance of the men. This, indeed, particularly since the payment of the prize money, has been carried to a degree of excess, which I had never before seen; and exceeds even credibility. When frequent intoxication happens in a healthy ship, it does not produce any lasting bad effect; but if it occurs where the contagion of fever is present, it is a never-failing cause of an increase of the disease. From what I have observed, I believe it to be the most powerful remote cause; and I have no doubt, that the difficulty in conquering the contagion, at so favourable a season of the year, is in a great measure to be ascribed to excessive intemperance.

(Signed)

Valiant, July 20, 1794.   Geo. M'Callum, Surgeon."

especial manner, the action of contagion and marsh miasma.

I have often remarked in the wards of Haslar, during cold weather, that weakly patients, just recovered from fever, frequently relapsed, by getting up, though they did not go out of doors. The allowance of coals at that time was too small for a large ward: besides, the body thus debilitated, was very sensible of cold air. We also see instances of relapse by a person going too soon out of doors, or a sailor, on board, exposing himself too soon on deck. The convalescent among the Portugueze, on board the Europe, by being exposed to the cold under the half deck, where they lodged, were very subject to relapses.

But I have known cases, of seamen while washing decks, seized with the rigors and shivering, which usher in the attack of typhus, after being exposed to the contagion in a sick birth. One of the most obstinate intermittents which occurred in the Vengeance, began in a landman, during the operation of scrubbing in a tub of cold water: this had been ordered by way of punishment for dirtiness. Now I apprehend the cold, in these cases, accelerated that accession of fever, which, for it, might have been delayed for a length of time.

We have a right to suppose, that typhus is a disease peculiar to cold countries, and that it is seldom found when the heat rises above 72° of

Faren-

Farenheit. If this is a fact, it will happen that the inhabitants of warm climates will be very liable to receive the infection when they come to northern latitudes; and more fo than the natives of thefe latitudes. It feems very generally admitted by phyficians, who have written on difeafes peculiar to certain regions or diftricts, that ftrangers are more apt to be affected than the refident inhabitants. The Yellow Fever of the Weft Indies, and the Remittents of India, with the Dyfentery of both, are moft fatal to Europeans newly arrived. The ague counties of England are deftructive to vifitors. It is proverbial of the farmers, in the unwholefome marfhes, marrying wives from diftant parts, as a fpecies of traffic to accumulate wealth; becaufe they know that a ftranger cannot live long in thefe fens.

When a ftudent at Edinburgh, I have often remarked the frequency of typhus fever among gentlemen from the fouth; but particularly thofe from the fouthern provinces of America. During the laft winter I ftudied there, from my intimacy with fome gentlemen from that country, I could not help taking notice of the frequent occurrences. Mr. Quin died that feafon, 1784-5; Doctors Gibbon and Lyon, with fome others, recovered with difficulty.

A Portugueze fquadron came to Spithead in 1793; and became fo fickly, that they were obliged to land their men at Haflar Hofpital: many

of

of them died with fymptoms of great malignity. The fate of the fquadron which we have already mentioned, was ftill worfe; yet both left Lifbon in their ufual health. We know that natives of Africa are very liable to our fevers; in fhips they are commohly amongft the firft fufferers.

We can, therefore, eafily fuppofe that the inhabitant of a warm country, accuftomed to the ftimulus of the fun's heat, from 75° to 85° of Farenheit, by changing to a northern latitude, will be more liable to difeafes which originate from deficient excitement. By withdrawing the cuftomary ftimuli, the body is left in a condition more eafily to be acted on by other impreffions; hence the effect of contagion: in other words, they have all that has been called pre-difpofition to favour its action *.

What are called the active paffions, or fomething that ftrongly engages the attention, have been reckoned among the prefervatives againft in-

* It may be in the recollection of fome of my Readers, that the combined Fleet of French and Spaniards, when off Plymouth, in fummer 1779, which was my firft cruize at fea, were over-run with a contagious fever, which made them leave the Channel. It is probable, that every Spanifh Fleet will fhare this fate on our coafts. But there are not more than four months in the year, that they could venture up Channel; and at a time when we are dreading invafion, it may be ufeful to make the moft of this fact.

fection.

fection. We have always thought it of confequence to engage the minds of a fhip's company
when expofed to a fever, and when it cannot be
done by the neceffary duty, fomething elfe fhould
be fubftituted. To this head ought to be referred fome of the means of prevention which have
been devifed at different times, fuch as charms,
amulets, &c. The vinegar of the four thieves has
probably been of ufe in this manner.

It is in the abfence of ftronger impreffions that
fear feems to favour the action of contagion;
hence imaginary fears of taking the fever are fometimes converted into reality. I have feen thefe
apprehenfions fo ftrong, that men have been fent
from fhips as if under the immediate attack of
the difeafe, becaufe they anfwered every queftion
put to them, concerning the fymptoms, in the
affirmative. The confequence of this was, that
they were put into a fever ward at the hofpital,
and there got the genuine complaint.

We have been informed that perfons labouring
under a fit of the gout efcaped the plague; and
we have a right to conclude, that it would have
the fame effect in this fever. But typhus itfelf
will fufpend fome affections of the body. I have
known the virulent gonorrhœa difappear for a
fortnight, and return with the fame violent difcharge, and *ardor urinæ*, as when the fever firft
came on.

I will

- I will mention an occurrence, which came within my own obfervation, and which fhewed that ready recollection that, fo eminently diftinguifhed the medical practice of the late Dr. Cullen. In the houfe where I lodged in Edinburgh, my friend, Dr. Gibbons, lay ill of a fever, and was attended by Dr. Cullen. The miftrefs of the houfe, having overheard that part of our converfation which related to the manner in, which Dr. Gibbons had received the infection, was immediately impreffed with the idea that fhe would foon be feized likewife. Her fears operated fo ftrongly, that fhe faid, next morning, fhe had now got the fever, and determined to confult Dr. Cullen. Watching the Doctor with great earneftnefs, as he came out of the fick chamber, fhe put her wrift to his hand, and begged him to feel her pulfe, for fhe was very ill indeed. The Doctor, with much feeming indifference, walked paft, and did not even bid her a good morning. Such treatment, to a lady juft feized with a contagious fever, was beyond all endurance. She inftantly reddened, and grew angry, exclaimed with great vehemence againft courtly phyficians, and muttered fomething about large fees, &c. However, during this increafed excitement from the fuppofed affront, the fever vanifhed. It broke the chain of ideas that revited the imagination, and

like

like a faithful Archæus, faw the fpot which was
difordered, and corrected it.   Dr. Cullen, on
vifiting his patient next morning, with his ufual
complaifance, afked the lady how her fever went
on ? but fhe was now aware of what the Doctor
had intended, and confeffed that fhe was cured of
her apprehenfions.

I think I have feen means, fimilar to this conduct of Dr. Cullen, employed with much feeming
fuccefs, on the firft fufpicion of infection, among
timid people.

It has been long a general remark in our Navy,
whenever contagion appears, that its earlieft and
moft certain attacks are among the raw landmen.
Thefe, as being ftrangers to the modes and habits
of a fea-life, are for fome time aukward and dif-
pirited, and do not at once get into the method of
keeping themfelves perfectly clean.   It was obferved in the Vengeance, that new men who came
to fill up the places of thofe that went to the hofpital, were generally but a few days on board till
fome of them would complain.   Draughts of this
kind were received in the fhip three or four times,
while fhe lay at Spithead; and always with a repetition of this circumftance, though much attention was paid to cleanlinefs.   The fame thing
happened in the Valiant, and it feemed to prolong
the infection.

                                    When

When infection has been received into a ship, sometime in commission, although the usual proportion of landmen is on board, it has been found to be easily checked, and sooner expelled, than otherwise.

We have more than once seen a fever appear among the marines, who are berthed together, without affecting a single seaman for some time; and if timely attention was directed to the means of prevention, it might be overcome at the first.

In January 1790, the Gorgon, of forty-four guns, arrived at Portsmouth, with a number of troops on board, from Chatham, called The New South Wales Rangers, under the command of Colonel Grose; to which place the ship was bound. While she lay in the harbour, a fever prevailed among the soldiers, and several died. Accounts were brought to Admiral Roddam, the Commander in Chief, that the mortality was considerable, and the disease very dangerous. He ordered me to visit the Gorgon, and report to him the condition of the troops, and the nature of the sickness. I found fifteen or sixteen men ill of a typhus fever, and one or two with symptoms of immediate dissolution. Upon careful investigation, and inquiries put to the convalescents that were walking about, I found that the fever had been brought into the barracks at Chatham,

where

where they were embodied, by two deserters from the guards, who came from the Savoy prison, and entered into the corps, on condition of pardon. Had these circumstances been attended to at first, this infection might have gone no further. The sick lay in the great cabin of a hulk, along side the Gorgon, which I requested the Admiral to order to be moved to some distance. This being done, every person with the first symptoms of indisposition were separated, and all communication was prevented. The Lords Commissioners of Admiralty, on receiving my report, were pleased to order me to superintend the people ; and every thing necessary to the comfort and recovery of the soldiers, which I thought proper to prescribe, was complied with. In a fortnight from this time, no fresh attacks were known ; a few died, but the whole afterwards did well. What I am about to draw from this fever in the Gorgon, is the circumstance of the seamen in the ship being totally exempt from it : not one of them having had the slightest sign of infection. Now the troops lived in the lower deck ; and when the ports were shut at night, the exhalations from below naturally ascended through the gratings, among the sailors.

In the Vengeance, very few of the able seamen were affected. But her condition was particular ; she had on board not only her own complement of landmen, but a number she was carrying round to

the

the Fleet. The infected were alfo fome time in the ſhip; and ſo late as the 12th of February, when I joined her, eleven were ill in the ſick birth; this, with the feaſon of the year, tended to prolong the contagion.

When the Valiant received near four hundred of the London's complement, ſhe had on board two hundred in perfect health; yet to the London's people was the infection for a long time confined. On the day before ſhe ſailed with the Channel Fleet, a landman came on board as a volunteer. The attacks of fever, for fome time paſt, had been very gentle; yet this man received the infection in a few days, and died at ſea. He had been brought up to a country life, and by being a ſtranger here, and not under the impreſſions of accuſtomed ſtimuli, ſoon fell a victim to the contagion.

In the Cæſar and Leviathan, our remark was fully verified. The former ſhip received from the London about ſeventy men; and the latter the whole of her marines. The Cæſar had on board the crew of the Ganges, who had been all turned over, after being twelve months in commiſſion. The Leviathan had been in the Mediterranean, and her people were healthy and ſeaſoned: both ſhips were under excellent diſcipline. In the Cæſar the fever did not extend above thirty caſes; theſe, according to Mr. Seed's report, were almoſt

confined

confined to the men of the London. Of the
Leviathan's, thofe fent to Haflar, were chiefly
marines, and but few feamen. Yet thefe fhips
were quickly cleared, even although the Cæfar
kept fome on board for cure.

This circumftance, of fhips well difciplined,
and long feafoned to fervice, being fo eafily cleared
and fecured from contagion, was certainly one of
the great caufes of fo many fhips in the Fleet
being quickly free of the fever communicated from
the French prifoners.

In a former part of this work, I have mentioned,
that while at Haflar, fome very bad cafes of typhus
were fent from the Coloffus and Robuft, in Portfmouth Harbour, both of which had lately returned
from Toulon. Yet the infection made little progrefs, and was quickly overcome. I do not think,
that a fhip's company, juft raifed and mixed,
ftrangers to their officers and one another, and
under no eftablifhed difcipline, could have efcaped
on thefe terms of health. It was, moreover, the
winter feafon, and by coming lately from a warmer
climate, they were the more fufceptible of typhus
contagion.

How

How CONTAGION *is taken into the Body.*

IT is a point generally admitted among Phyficians, that the atmofphere is the *medium* of Contagion, whether it arifes directly from the fick perfon, or from cloathes and other articles that have been impregnated with the noxious matter. We have mentioned certain conditions, by which this matter may be rendered more powerful in its operation, by being collected, or more concentrated; and alfo certain conditions of the found body, that favour its reception more at one time than at another.

On approaching the fick bed of a patient labouring under a malignant typhus, if the room is fmall, and not well ventilated, the fmell which we have formerly mentioned, becomes very fenfible. If the patient fuddenly ftarts up in bed, or haftily throws the bed-cloaths from him, it is ftill more perceptible. To people not accuftomed to vifits of this nature, the air becomes particularly offenfive; a difagreeable tafte is excited in the mouth, and frequent fpitting takes place: unpleafant ideas at the fame time are formed, and affociated with the recollection of the ghaftly looks of the fick man, which cling, as it were, to the memory. If the mind attends feelingly to thefe fenfations, they lead to others which evidently fhow the effects of contagion; the change of air does not feem to alter the nature of the fmell; and the tafte of the

P 3

mouth

mouth is rather more difagreeable, with drynefs,
even to a flight degree of naufea, and fome efforts
to vomit. A langour, lownefs of fpirits, inaptitude
to motion, apprehenfion of evil, drawing near the
fire, and drooping the head, with a fenfe of
heavinefs of the eyes and forehead, follow next.
A creeping fenfation is felt over the fkin, which
becomes paler and dryer than ufual: the coldnefs increafes, with now and then a flight fhivering: there is alfo a degree of anxiety, with pains
of the fmall of the back, and limbs, or ftitches in
the fide and breaft, as they are often called. The
coldnefs on the furface and fhivering, fometimes
ftop for a while; or are alternated with a tranfient
glow of warmth, which foon vanifhes, when the
rigours and fhaking or trembling recur with
more violence: even to make the patient grafp at
fomething to fupport him, left he fall down.
The countenance now appears more dejected and
dull, and the features fhrink, the face is fometimes
fuffufed with tears, and now a confufion of
thoughts takes place. The heart is occafionally
affected with palpitation; and the pulfe flutter-
ing and feeble. Women fall into hyfteric fits;
and I have even feen men who had the affection
of the *globus hyftericus*, as exquifite as in the fair
fex. I have alfo feen epileptic paroxyfms at the
acceffion of typhus.

Contagion can be often traced from its fource,
to its firft reception into the body, in this manner,

till

till the fever appears in its full form and character. From cloathing and other fubftances imbued with the exhalations from a fever fubject, the fame train of fymptoms follow. Seamen are very often infected in the act of tying the bedding and cloathes of their fick mefs-mates, when they go on fhore to hofpitals.

The period for the fever, or fymptoms of infection, making their appearance, varies according to the concentrated ftate of the contagious matter; or the caufes and circumftances which favour its action. We have juft traced it from the moft early effect; but at other times, days and even weeks are neceffary to produce thefe appearances in the body.

The difagreeable tafte of the mouth, the fpitting, naufea and vomiting being fuch early fymptoms, have induced fome to conclude, that the infectious matter is received by the faliva, taken into the ftomach, and from thence by the circulation conveyed to the body at large. This way, however, does not appear to me to be the mode of its action: for by throwing out the faliva, we fhould be fecure againft it; which we know not to be the cafe. The effects of the poifon are fometimes fo fudden, as to give no time for this circuitous mode of operation. While the people were unpacking the bale-goods at Marfeilles, they were affected inftantly: we have feen delirium in many perfons, the firft fymptom

of typhus; we muft therefore fuppofe, that it
acts immediately on the powers of life. We can
fee no objection, why it is not communicated to
the nervous fyftem thrqugh the lungs, independent of any chemical change, which an impure
atmofphere may effect in the blood. As the air
conveys the hurtful quality, it is furely more reafonable to think that it acts by the organs of refpiration, than by the ftomach or lacteals. The
ficknefs, and anorexia, appear to me fymptoms of
mere confent of parts, which conftitute a fhare of
the phænomena of fever, and only fecondary.

A certain quanity of infectious matter, it would
appear, is neceffary to affect the body with difeafe.
But I confider the fymptoms of fever, as motions
of our fenfitive fyftem; excited by impreffions of
the noxious power, that are prejudical to health,
and not as efforts of a *vis medicatrix naturæ*; they
are only figns, that the prefent ftate differs from
the cuftomary tenor of our habits and propenfi-
ties. Now the phænomena of fever, being attendants on very different conditions of body, we
have a right to conclude that they are *always*
fymptomatic.

If the impreffions made on our fyftem by the
contagious matter, whether breathed or fwallow-
ed, have not been fufficiently ftrong, they induce,
either a trifling indifpofition, or none. But if they
have been ftrong enough to produce difeafe; and
if the means made ufe of to overcome them *in*

limine

limine, have not had the defired effect; it beomes
a difficult matter to eradicate them afterwards.
They feem to occupy the whole attention of the
body; their prefence conftitutes the length of
the fever. If they wholly fupprefs the accuftomed actions and habits which we exercife, it is
death. But if the accuftomed motions and actions, which by length of time had been confirmed
into habit, fhould gradually recover their former
energy, then the difeafed ftate declines in proportion, till the body at laft returns to the vigorous enjoyment of all its appetites, which is *health*.
We know nothing of what has ufually been called
a crifis; and far lefs of concoction and evacuation
of morbific matter.

Treatment of Infected Perfons.

IF after a perfon has been expofed to contagion, few or more of the fymptoms which have
been enumerated take place, there will be reafon
to expect a fever as the confequence.

It would feem to be the practice of the beft
phyficians, to prefcribe for the ftate of the patient, not with a view to expel the contagious
matter from the ftomach or body, but to counteract its effects. The bad tafte of the mouth,
the want of appetite, and ficknefs at ftomach,
very naturally indicate an emetic; and this has

been

been the moft general practice. About fifteen
grains of ipecacuenha, joined to fome of the
milder antimonial preparations, will anfwer the
purpofe : or, the following formula :

> R̵ Tinct. ipecac. - - ʒi.
> Vin. Antimon. - - ʒiij.
> Ft. Hauftus.

The beft time of exhibition is towards the
evening. The patient ought to be put to bed ;
and at bed-time, let him take the following bolus
or draught :

> R̵ Pulv. Antimon. - - - gr. v.
> Opii purif. - - - - - - gr. i.
> Confect. Aromat. - - Ft. Bolus ;
> Hora fomni fumendus: *Or*

> R̵ Vin. Antimon. - - - ʒij.
> Tinct. Opii. - - - gᵗ· xxvij.
> Aq. Cinnam. - - - - ʒi.
> Ft. Hauftus; h. f. f.

A few glaffes of wine made into negus, ftronger
or weaker according to the perfon's habits of
drinking, or warm wine and fpices, may be
taken afterwards ; and then let him be left to his
repofe.

In the morning, if the bowels are open fo much
the better ; but otherwife, a gentle dofe of the
tartarized infufion of fenna, will be ufeful. Though
it may be neceffary to keep the patient quiet ;
yet his mind fhould be fo employed and amufed,

as

as to divert him from all unpleafant feelings. The air of the room fhould be pure, but it is neceffary at the fame time, now, and throughout his illnefs, that it fhould be fufficiently warm. I think it ought never to be below 60° of Farenheit, or above 68; and often changed. If the weather out of doors permits going abroad, I would recommend it.

It is at this period of typhus only, that preparations of antimony appear to be ufeful. I know not how they act; I confider them as lefs active ftimulants, when given in fmall dofes. The pulvis antimonialis ought therefore to be continued, to the quantity of four or five grains, three or four times a day ; and the opiate combined with the laft dofe at bed-time : wine may be given in proportion to the debility and defire for it, fo as to fupport the pulfe ; which fhould be carefully watched, as by it, we muft regulate the ftimulant plan, without carrying it to excefs, or inducing that debility we wifh to avoid. When our patient is thought to be free of all complaints but weaknefs, let moderate dofes of cinchona be given, and let him keep warm, ufe exercife, amufement, and a diet nourifhing and ftimulant.

How

How to subdue Contagion.

If pneumatic chemiſtry has aſſiſted us in ex-
plaining the nature of contagion, and the manner
it is conveyed from one perſon to another; it alſo
teaches us how to weaken its power, ſo as to be
no longer dreaded.

The great experience which the late Dr. Lind
had on this ſubject, in an hoſpital, better calculated to afford information than any other in Europe, juſtly entitled his opinions to be heard with
confidence, and practiſed univerſally. On a
branch of medicine, however, always traced with
difficulty, and ſcarcely to be known but by its
effects, conjecture ſometimes obtruded inſtead of
proof, and haſty concluſion in the place of de-
monſtration; hence we find this celebrated phyſician aſſuming a fanciful hypotheſis for his
theory, and on which he builds the whole ſuper-
ſtructure of his practice. The following paragraph
may be taken as his text: he ſays, " The cleareſt
" idea we can conceive of the manner in which
" this infection is communicated, is to ſuppoſe,
" there is in all infected places, adhering to cer-
" tain ſubſtances, an envenomed *nidus*, or ſource
" of effluvia, *corpuſcules* or whatſoever infection may
" be ſuppoſed to conſiſt, and that as the air is more
" or leſs confined, becomes more or leſs ſtrongly
" impregnated with them *." In ſome parts of

* Papers on Infection.

his

his valuable papers on fever and infection, the Doctor appears to be satisfied that *animalcula* had nothing to do with the propagation of contagion; but here he relapses into the old doctrine, and supports it in new terms, and an improved nomenclature. He looked upon infected cloathing to be this envenomed *nidus*, where a new generation of *corpuscules* were multiplied, which gave to the noxious matter an increased virulence, and became more certainly productive of the disease, than when it first issued from the body. That this was his meaning, is abundantly confirmed by all his rules for expelling and purifying contagion. Every substance which he prescribes is with the direct intention to destroy animal life; hence the heat of an oven, fumigations with brimstone, gunpowder, tobacco smoke, boiling vinegar, &c. are his executioners. It was not only to purify cloathes, bedding, and other substances, that these fumigations were applied, but the wards of the sick in hospitals, and the decks of ships, became also subjects for the process.

Heat I consider as one of the most powerful correctors of contagion; it rarifies foul air, or what may be spoiled by respiration in crowded apartments: applied to substances imbued with animal miasma, it will dissipate that, or convert it into an inert mass, so as to be harmless. Wherever there is moisture, it will dry it; and above all,

it

it is ufeful, as a general ftimulus to the body, keeping it warm, and thus fortifying it againft cold, which fo evidently tends to difpofe it to receive infection.

Of fome of the fubftances ufed in the fumigating procefs, I have nothing to fay in praife. I do not think that they are harmlefs only, but hurtful. Under the high authority of Dr. Lind's name, they are thought to be the only fafe and certain means of expelling infection, and have been long practifed in our fhips *. When this is to be done in an infected fhip, but it is often performed in health as a preventive, the people's cloathes, bags, and hammocks, are all hung up or fpread abroad in the lower deck, in order to be prefectly impregnated with the fumes; the fcuttles and ports are clofely barred in, and the tarpaulins clofely laid over, that nothing may efcape. Gun-powder, brimftone, &c. are then burnt below in quantity fufficient to pervade every corner. This is fometimes fo compleat, as to kill cats, rats, and mice, or other vermin in the fhip. During this procefs, which is occafionally continued daily for two or three hours, the fhip's company is kept aloft; not always in the moft temperate weather.

* In the Colloquia Maritima, publifhed in 1688, I find that fumigations were employed to correct foul air : fee quotations from this book, under the article YELLOW FEVER.

Some

Some officers have a high opinion of the fumigations; yet I have been often afked, in infected fhips, how it came that the fever was ftill spreading, though they perfifted fo carefully in fmoaking? The truth was, other means became fecondary confiderations; fo great was the faith in the burning brimftone, that little attention was paid to the early feparation of the difeafed. But this was not all; the people being expofed to the cold on deck, in a ftate of inaction, were evidently hurt by it. Immediately after the procefs was finifhed, frefh cafes made their complaints known, and it feemed to haften the rigours and fhivering. I have feen this practice of fmoaking continued without intermiffion for weeks; and when I could not be prejudiced againft it, have found the contagion go through its regular courfe, and at laft die away of itfelf. In the winter feafon, when there is danger of the feamen fuffering from the inclemency of the weather, I have conftantly forbid it. I know no condition of a well-regulated fhip, that can ftand in need of thefe vapours.

When a fever of this kind breaks out in a fhip at fea, particularly in bad weather, fhe muft of neceffity quit the fea: but a fick birth fhould be immediately fitted for the infected, as far as poffible from the others, and all communication ftrictly prevented. A man of war admits of few conveniencies in fuch a cafe; but the Captain,

under

under such circumstances, should have it in his power to order slops and bedding, when wanted, from the stock of the purser, that the sick may be often shifted; and soap should be allowed, to keep every thing clean that belongs to them.   Much might be done by attention to cleanliness, in both cloathes and person, and it would be a great means of checking infection on the earliest discovery. Some steady old men, who run little risk of receiving the disease, should be appointed attendants on the sick, and furnished with every thing neceffary; such as a few coarse cloths for wiping the patients after being washed every morning; spare night-caps, after the hair has been cut, and shirts and frocks made into sheets *.

On suspicious sickly men coming into a ship, the surgeon ought to examine them attentively; about the places they came from; whether any fevers or other diforders prevailed in the neighbourhood, and also in what vessels they were conveyed from the out-ports to the ship.   An in-

* I have been told, that the new Medical Board of the Navy, have it in contemplation to supply every ship with a certain quantity of bedding. I would recommend the mattrasses to be covered with leather, or canvas painted, so as to exclude moisture from the hair within ; by which means they would neither imbibe infection, or be difficult to be cleaned. This method of constructing mattrasses would be a great improvement in all hospitals. With these articles we must expect soap, coarse towels, &c.

fected

fected subject ought to be sent out of the ship with all speed, and every thing belonging with him. There is no certainty in any means of expelling contagion, but removing the infected. The slightest case should not be left on board for some time, till there is an assurance that the strength of it is subdued. I have seen a man kept in the ship, because he was thought to have the fever in a milder degree, and being a useful man, which has prolonged the infection for weeks. There ought, therefore, to be no deviation from this rule. In the present improved discipline of our Navy, I have never found it a very difficult business to eradicate an infectious disease, provided my directions were duly complied with. The system of duty in our ships at this moment, is so perfect in all its branches, that I have nothing to add. The whole preservative means are comprised in, the immediate removal of the sick, cleanliness in person and cloathing; fires, to keep the people, in the winter season, warm; avoiding cold and moisture, fatigue and intoxication, and keeping the ship dry, and properly ventilated. In an infected ship, an active and sensible officer will be employed in airing his people's bedding and cloathes; distributing orders to the inferior officers, to see that their divisions of the seamen are clean in their persons, that their dirty things are washed twice a week, and that they have cloathes sufficient to keep them warm. If the weather is cold, we

Q

shall

shall see him ordering the decks to be dry rubbed
with sand, or scrapers, and washing with water laid
aside; his men will not be allowed to go aloft
when it rains, or into boats when it blows hard.
To give the decks a cleanly appearance, they will
be well white-washed fore and aft, above and
below: and lastly, to combine in his method the
advantages of pure air with warmth, the ports will
be opened to leeward, and only the scuttles to
windward; or the whole fitted with sashes of
bunting, and stoves lighted in every part that can
do good.   To all these means of preserving health,
amusements will be found to keep the minds of
the people in action; violins, and other instru-
ments of music, being common in most of the
King's ships, are usually employed in the evening,
and the seamen and landmen are seen joined in
the dance*. A physician of a fleet, who may visit
a ship

* It has often occurred to me, that a band of music would
be extremely useful in a ship, even as a preservative of health.
In the last ship where I was surgeon, the Vengeance, Captain
C. Thompson, and the Officers, purchased musical instru-
ments; and five or six men, who were performers, made into
a tolerable band.   The people were regularly piped to dancing
every evening: and I always thought it but justice to allow it
a share of credit in the extinction of the typhoid contagion.
To a set of human beings, confined for months together within
the gloomy walls of a ship, the exhilarating powers of music
could not fail to produce the most salutary effects.   Our immortal Shakespeare very justly thought that man fit *for crimes*,
who could not be touched with the concord of *soft sounds*.

a ſhip under the command of an officer of this deſcription, though·armed with a diploma, and with the chemiſtry of the elements at his fingers ends, will find that very little has ·been left for him to do; whether his doctrine of prophylactics, be the *vinegar of ·the four thieves*, or the fumigations of modern phyſicians, under the ſcientific appellations of Sulphurous Gas, Muriatic Acid Gas, or Nitrous Gas.

" Spithead, July 2, 1796.

" SIR,

" THE ſubject of a contagious Fever, on board the Niger frigate, at Spithead, having attracted the attention of the Lords Commiſſioners of Admiralty, to order a particular proceſs for the purification of the ſhip; I have to requeſt you will be pleaſed to communicate theſe remarks to their Lordſhips, as I conſider the welfare of the Navy deeply intereſted in the diſcuſſion.

In like manner I would pronounce a phyſician ſtrongly fettered in the craft of *technicals*, that excludes from his *Hygeine*, the exciting influence of melodious notes, or the agile movements on the " *light fantaſtic toe*." I am myſelf paſſionately fond of both; and nothing gives me more pleaſure than to ſee it extended to the ſeamen.—This naturally makes me aſk, why a regiment is allowed a band more than a ſhip of the line?— Give your Tars, O my countrymen! their amuſements; and while you enjoy your's on ſhore, remember who they are that give them ſecurity!

Q 2

" Im-

" Immediately on being informed of the Niger's situation, I could not help thinking that the fever had been brought from some of the vessels, which were lately boarded on the coast of France. Captain Foote since tells me, that some bedding had come to the Niger from a brig that had been employed in carrying troops from Brest to Bayonne.

" In reporting the condition of the Niger to Vice-Admiral Colpoys, I considered it necessary that she should be detained, till the disposition of the contagion should be sufficiently known, or so far subdued, as not to endanger the lives of the people. From the cases that were examined, I dreaded no mortality; and from the compleat mode of discipline established under Captain Foote and his officers, with the careful separation of the infected, I did not doubt of the speedy extinction of the fever. They had judiciously practised every means that have been found most successful: and a concern for the health of his people, truly paternal, has directed Captain Foote's endeavours on this occasion.

" A pamphlet containing Dr. Smyth's address to Earl Spencer, has just been put into my hands. I now conceive it to be my duty to make some observations on the manner in which it seems to have engaged their Lordship's attention*.

" There

* " An Account of the Experiment made at the desire of the Lords Commissioners of the Admiralty, on board the Union

" There can be but one opinion, that pure air is fitteft for animal life : therefore the correction and expulfion of a tainted atmofphere in fhips, have always employed a fenfible officer.   Fires are occafionally put into the holds and wells, to dif-lodge *carbonic acid gas*, or in plain Englifh, fixed air.   Another foul air, our officers are in the habit of expelling, is *azote* or *mephitic air :* this is per-formed by opening ports and fcuttles, unlaying gratings, and putting down wind-fails.   The fixed air of the well, is produced by the decompofition of water and vegetable matter, as the timbers of the fhip.   The azote is of animal origin, and abounds, whenever the air is polluted, by breathing animals ; as, between the decks of a fhip.   It de-rives its name from being fatal to animated nature. Now this azote is the bafe of nitrous acid : they only differ in the degrees of combination with oxygene, or what was formerly called dephlogif-ticated air : and in proportion to the quantity it attracts of this principle, it is called Azote, Azotic Gas, Nitrous Gas, Nitrous Acid, Nitric Acid.   In fhort, Dr. Smyth's Preventive, is the very fub-ftance that every intelligent officer is hourly em-ployed to drive from the decks of his Majefty's ships.

" In

Union hofpital fhip, to determine the effect of Nitrous Acid, in deftroying Contagion ; by J. E. Smyth, M. D. F. R. S. &c. *Johnfon, London,* 1796." -

" In Doctor Smyth's procefs, when the nitrous acid is converted into gas, it lofes a portion of pure air ; it is now an elaftic fluid, under the title of Nitrous Air, or Gas ; in this ftate it will remain for fome time, till it again, by chemical attraction, recovers its pure air : which it will do fooner in proportion to the purity of the atmofphere, when by its fpecific gravity it will fall to the deck, Nitrous Acid ; or to be more intelligible, it is turned to *aqua fortis* ; for the nitrous gas is nothing elfe but the fumes of aqua fortis. Azote, the bafe of the nitrous gas, which I am now fpeaking of, being of an animal nature, is produced in greateft quantity, during the putrefaction or decompofition of animal matter : by expofure to the air it paffes from azote, through all the degrees, till it becomes nitric acid ; when by throwing pot-afhes to it, it becomes falt-petre ; witnefs the gun-powder fyftem of the French democracy, where all animal fubftances have been put in requifition for the formation of nitre. Had Captain Foote, of the Niger, inftead of duly ventilating his fhip, with a change of air, fhut every port and fcuttle, laid the tarpaulins clofely over, and ordered every man below, an atmofphere of azotic and nitrous gas would have been formed, in fuch quantity, that would have fuffocated his whole crew in the fpace of a few hours.

" If what I have afferted be matter of fact, how improper muft it be to introduce this lethalic

vapour

vapour between a fhip's decks, that ought to be occupied with pure atmofpheric air. The conclufion to be drawn from it, does not reft on my authority; it is well known to every chemift.

" The idea that has given birth to thefe experiments, is evidently a remnant of the exploded doctrine, that contagion originated from *animalcula*. Hence a diftorted philofophy ordered us to burn brimftone, and fire gun-powder in our fhips: its Protean form, under the nitrous gas, is equally repugnant to chemical truths, as it was before. If fuch has not been their origin, there is an unpardonable negligence in chemical fubjects, in a work avowedly held out as confirmed by trial and experiment.

Their Lordfhips have been told, that the bad fmell difappears from the ufe of the gas: but any other fubftance, as the fmoke of tobacco, or afafœtida, that gives a ftronger impreffion to the olfactory nerves, will do the fame thing.—But then it will be faid, the fuccefs on board the Union hofpital fhip, proves its utility. I flatly deny, that this appears from the narrative itfelf; there is a deception, from firft to laft, in the bufinefs. Among whatever body of men contagion makes its appearance, there will always be found fome more fufceptible of it than others. Thofe who are moft difpofed from particular caufes to receive infection, will be firft feized; it will then extend to others; and there will, at last, be found some who escape it altogether. Even the plague itself proceeds in this manner. Now, if any *charm*, *nostrum*, or pretended preventive, be introduced at the stage of infection, as happened in the hospital ship, Union, it will gain a credit it never merited; and the exemption from the disease ought to be explained by a very different mode of reasoning. This has been the tenor of all prophylactic processes, from the vinegar of the four thieves, to the present invention. I might produce a numerous train of facts to support further arguments; but they cannot be made sufficiently intelligible to those inaccustomed with medical and chemical disputation. They will appear when the Medical Records of the Channel Fleet are published, with a mass of evidence, on the subject of Contagion, greater than has usually fallen to one observer. If success ought to give preference to modes of practice, I might appeal to the sudden extinction of the fever, spread from the French prisoners after the memorable 1st of June 1794, to two thirds of our ships; and in more than thirty other ships that have been cleared under my directions at different times.

" It ought to be remembered, that in a clean and well-aired ship, contagion can lodge no where but in the bodies, cloaths, or bedding of the people.

people. When we purify, by wafhing and airing all thefe articles, or remove them, the fources of contagion are deftroyed. Nay, whatever renders the air more unfit for refpiration, muft alfo difpofe the body more to receive infection; becaufe it weakens the powers of life. I even confider the moft agreeable aromatic odours fuperfluous, becaufe they may deceive our fenfes of fmell, with refpect to the ftate of the atmofphere; which, when pure, and frequently changed, admits of no foreign aid. By thefe means, the exhalations from infected bodies become diluted by fo large a portion of air, as no longer to convey difeafe*.

" Having faid fo much, I confider myfelf bound to explain any obfcurity which may appear, fhould

---

* Air flues, or pipes, might certainly be conftructed, on fome plan, to facilitate the efcape of the foul air from the decks where the people fleep. When five or fix hundred men lay on one deck, the fpace cannot be always fufficiently ventilated by the hatches; for if it happens to rain, thefe are occafionally covered with tarpaulins; fo that the pure part below muft be rapidly confumed. Now the moft noxious portion of the air, from fpecific levity, occupies the higheft ftratum where the hammocks are flung: fmall tubes, to communicate with the air aloft, would therefore conduct the tainted atmofphere upwards, as quickly as it was collected; and would admit a frefh draught for a more healthy refpiration.—All hofpitals, and places where a number of people crowd together, fuch as affembly-rooms and theatres, ought to be furnifhed with thefe conductors for the efcape of azotic gas.

their

their Lordſhips think proper to ſubmit my opinions to any authorities competent to decide.
They will forgive my intruſion on this buſineſs,
while it remains a part of my duty to infuſe into
our ſyſtem of health, every diſcovery which the
auxiliary branches of medical ſcience may pub-
liſh; while, at the ſame time, I ought to take
care that nothing noxious may enter either the air
or diet of a ſeaman *.

I have the honour to be,
SIR,
Your faithful humble ſervant,

To Evan Nepean, Eſq.   T. Trotter."
Admiralty."

The Lords Commiſſioners of Admiralty were
pleaſed to order their ſecretary to tranſmit the
above letter to the Commiſſioners for ſick and
wounded,

* I am obliged to Mr. Burd, ſurgeon of the Niger, for the
copy of my original letter; which he took as it was about to
be ſent to the poſt; otherwiſe, I ſhould not have been able
to give it a place here, having had no time to tranſcribe it.—
It was my wiſh, that the Lords Commiſſioners of Admiralty
ſhould be acquainted with my opinion, of this proceſs for
deſtroying contagion, before the health of the Niger's people
could be aſcertained. By this I muſt beg leave to acquit
myſelf of any miſconſtruction or pre-poſſeſſion of Dr. Smyth's
form.

wounded, for their report; which is contained in
the following letter to Mr. Nepean, a copy of
which was tranfmitted to me.

" Office for Sick and Wounded Seamen,
July 11, 1796.

" S I R,

" We have received your letter of the 5th inftant, inclofing one which you had received from
Dr. Trotter, phyfician to the Fleet, and fignifying
the directions of the Right Hon. the Lords Commiffioners of Admiralty, that we fhould confider
and report to you, for their information, our opinion on the feveral matters ftated by the Doctor,
on the fubject to which it relates.

" We obferve what we conceive to be inaccu-
racies in feveral parts of the Doctor's reafoning,
fome of which we fhall point out to their Lordfhips. He alledges, that the azote, as it is called
by the French chemifts, or phlogifticated air, as
it is called by the Englifh philofophers, differs
from the nitrous gas, and nitrous acid, only in
degree; whereas the former is underftood to be
intirely without vital air, which is one of the
component parts of nitrous gas and nitrous acid,
and by giving them the properties of acids, renders them effentially different from the pure azote.
We might farther remark, that this azote is fo far
from being deadly, or even noxious, except in its

moft

moſt concentrated ſtate, that near three fourths
of the air *which we breathe* is compoſed of it;
and that it is not correct to aſſert, that the azote
is derived from the living animal body.   For it is
conſidered as proved, in chemical philoſophy, that
the chief changes which common air undergoes,
when it is converted into the above mentioned
azote, by the reſpiration of animals, ariſes, not
from any addition it receives from animal effluvia,
but from the abſorption by the lungs of vital air;
which is the other principle component of the
atmoſphere, and which, when abſtracted from
common air, by breathing, leaves in a ſtate unfit
to ſupport animal life.   But we judge it unne-
ceſſary to detain their Lordſhips with further de-
tails of this kind, as we apprehend this queſtion
is to be decided, rather by plain experience, than
by ſcientific deduction. .

    " We agree with Dr. Trotter, as to the great
importance of attention to ventilation on board
his Majeſty's ſhips, together with perſonal cleanlineſs, which will, without doubt, prevent a poſſi-
bility of the generation of infection.   But, for the
ſpeedy deſtruction of infection already generated,
we are of opinion, that ſomething more than
ventilation is requiſite; and that the uſe of fire
and fumigations with ſulphur, is the moſt effectual
mode of annihilating infection.

                                            " With

" With refpect to the noxious quality of the fumes of nitrous acid, propofed to be employed to deftroy infection, by Dr. J. C. Smyth, we can produce unqueftionable authority that thefe fumes, when drawn into the lungs, undiluted with common air, are highly noxious. But in the diffufed ftate, in which they have been ufed by their Lordfhip's orders, it does not appear, that any confiderable inconvenience was experienced from them: and whether any bad effect would arife from thefe fumes, even when diluted, being applied to the lungs for a length of time, is a fact concerning which we have not yet had fufficient experience to decide.

" As to the nitrous fumes, in the manner directed to be employed by Dr. Smyth, being adequate to the deftruction of infection, which fact is controverted in fuch ftrong terms by Dr. Trotter, we have already intimated to their Lordfhips our opinion, that the trial on board the Union is not, alone, of weight fufficient to determine this matter; efpecially as the mortality among the Ruffians, and the fick among the attendants at Deal, where the ufual modes of prevention were practifed, was lefs than in the Union, where nitrous fumigation was put in practice.

" Befides the farther trials ordered to be made by their Lordfhips, we have directed one to be made at the Mill Prifon, at Plymouth, and fhall

continue

continue to embrace such fit opportunities as may offer, of bringing this matter to an unequivocal decifion.

(Signed)   R. Blair.

To Evan Nepean, Efq;   Gil. Blane."

Dr. Smyth's procefs for deftroying contagion, having been introduced into a fhip in the Fleet, it was not left for me to remain any longer a paffive fpectator of the bufinefs; otherwife I might have faved myfelf an unwelcome criticifm. It happened, however, that the fever in the Niger had been completely fubdued before the ufe of the nitrous gas; the credit of which was folely due to the attention of Captain Foote, his furgeon, and officers. Captain Foote was not a little hurt when the refident Commiffioner of the fick and wounded came on board this fhip, to detect the foul air which had caufed the fever. This gentleman was conducted into every corner of the Niger by the Captain; but the atmofphere proved to be fweet to the very keelfon; fhe was, indeed, a fhip in that kind of order, where a phyfician would always wifh to fix his field for experiment; for what depended upon the exertions of officers and men, would certainly be executed; and fo it proved with the expulfion of this contagion.

7

In

In addreffing the Admiralty on the fubject, with a defire to render my language familiar, I have been apparently betrayed into fome feeming inaccuracy, as mentioned by the Commiffioners of fick and wounded. I wifhed to imprefs their Lordfhips with the idea, that this procefs was really introducing an atmofphere into our fhips, which we are conftantly endeavouring to expel. There is no great difference between azotic gas and nitrous gas; and I contend, that the latter, however diluted by common air, is in direct oppofition to the purpofes we intend by a free ventilation. The formation of nitrous acid, is a procefs conftantly going on in nature; and whether we fubftract the oxygene, the vital air, or add nitrous gas, it is plain that we render the medium in which we breathe lefs fit to fupport life: hence I ufed the common word, *pollution*. Nay, a captain of a man of war would be acting up to the principle laid down, if he neglected all attention to opening ports and trimming windfails, to keep the gas in the decks as long as poffible. The definition of this gas is, that " *it extinguifhes light and deftroys animals.*" Now, to fay this becomes harmlefs, when in a diffufed ftate, is no argument in favour of its virtues; when refpired for a length of time, it may have all the bad effects that a fingle infpiration would have when it is undiluted. If a perfon takes a fmall draught

out

out of a quart of water, in which thirty grains of arfenic have been diffolved, he may feel no inconvenience ; but if he drinks more, or a little now and then, death may enfue.

There is one part of this queftion, that the advocates for this gas, or any other gas, as correctors of an infected atmofphere, have altogether overlooked. The difeafe that generates this infection, is the very offspring of an impure air : the very fymptoms characteriftic of it, fhow the deficiency of vital air in the body, mufcular debility, pale or bloated looks, and hæmorrhages of a darker blood. Some phyficians of eminence, at the head of them Dr. Beddoes, the celebrated chemical philofopher, contend, that the cure depends upon reftoring this loft principle ; and we all know how grateful a pure draught of air is to a typhus patient. Now while we attempt to preferve health, let us not forget our duty to the fick. But in the ardent purfuit after thefe factitious airs, to correct contagion round the bed of a fever patient, they feem to think of nothing elfe. Dr. Smyth, and his experimenters on board the Union, feem to have been fatisfied, if the fick men in bed did not cough from the irritation of the gas : nay, while one of them congratulates the Doctor on the immortal honour he will derive from his invention, and he in return, recommends him to the firft Lord of the Admiralty, both are confident that

the

the difeafed people received from it much benefit.
It appears from the report of the Commiffioners of
fick and wounded, that they are profecuting a final
determination, on the effects of the nitrous gas.
I do not think that it remains to be decided, but
is already done, by the united experiments of all
chemifts. They too, I find, from the opinion
they have given of fumigations of fulphur, are
fupporters of the old theory. If there had not
been a pre-conceived idea of the nature of contagion, how otherwife would phyficians have
ever thought of fuch agents as they have employed. Gun-powder and brimftone burnt by char-
choal, give out gaffes much refembling one another. The famous Morveau changed it for muriatic acid gas * : and Dr. Smyth now contends
for the fuperior efficacy of the nitrous gas. The
medicine has been changed, but the theory remains the fame; and what is a little ftrange, they
keep it in the back ground †.

Notwith-

---

* Morveau, the celebrated chemift : the laft accounts which
we have heard of him, was his employment as a naturalift,
capering after Bounaparte's army, in Italy. He has probably
been more fuccefsful in his new office; a fingular appendage
to an army!

† Suppofing, for a moment, that the exhalations of difeafed
fubjects, which convey infection, were *hydrogene*, whether combined with *fulphurous*, *phofphorous*, or *carbonic acid gas*; 'nitrous
gas,

R

Notwithſtanding the long and extended practice of fumigation in our ſhips, and at the hoſpitals,
it does not appear to me, that there are any clear
and deciſive teſtimony of certainty as to its effects.
The good effects of heat, I readily admit. But
heat is combined with the fumigation, in purifying cloathes and bedding at our hoſpitals. Probably, the heat is ſometimes not great enough,
from inattention in the labourers ; becauſe during
the period of my four months in office at Haſlar,

gas, according to elective attraction, would combine with the
oxygene; which it always does, and returns to its former
ſtate, nitrous acid, while the infectious matter would remain
the ſame. The experimenters on board the Union ought to
have told us, the preciſe condition of pre-diſpoſition, in all
thoſe people who had the fever from the infected Ruſſians:
their ages, conſtitutions, in what part of the hoſpital they did
their duty, whether they attended the worſt patients, and alſo
whether thoſe who eſcaped had not peculiarities of ſex, age,
temperament, &c. that reſiſted the contagion. In the extenſive practice which I have had in the Fleet, had it been my
intention to try any *preventive*, how eaſily might ſimilar proofs
have been laid before the public, to confirm my ſuppoſed ſuperior method, when it appears that a practice made familiar to
officers, is all the *mighty magic I have uſed*. I am not a little
ſurpriſed that Dr. Smyth, in compiling his report, entirely
overlooked the judicious alterations propoſed by Mr. Menzies: to which every reader, acquainted with medicine, will
readily beſtow more praiſe, than to the nitrous gas.—I ſincerely
wiſh that it may be the laſt effort of this kind, to draw the attention of our Public Boards from more valuable regulations to
ſubdue Contagion.

ſixteen

sixteen cases of typhus occured, that could be traced to the bedding, in chronic wards, which had been formerly in those of fever patients. But I 'have always thought it a fair conclusion, that two-thirds of our people, who relapse after returning· to their ships, owe this second illness to imperfectly-cleaned cloathes and bedding. In the last infected ship of the Fleet, the men returned with their cloathes all dirty. The Captain ordered them to be washed ; the men rose early to perform it, when seven were seized with convulsions and relapses so badly, as to be sent on shore a second time. Although I attributed the relapses of these men, to cold or fatigue, yet it justifies my general objections to this mode of purification ; and I long to see more attention paid to the cleanliness of our seamen's apparel, than has hitherto been done.

The following quotation shows that Dr. Lind himself, was not altogether satisfied with simple fumigations. He says, " But in prisons, ships, " or places, where the people cannot be moved, " and consequently a *sufficient* degree of heat can- " not be raised ; the application of fire and smoke, " to remove infection, may prove *ineffectual*. I " am confirmed in this assertion, by repeated " instances of infection in ships, both at sea and " when lying at Spithead, where every method " failed of putting a stop to it, as long as the

R 2 " men

" men remained on board *." It was to anfwer
the defideratum mentioned here, that Dr. Smyth
flew to the nitrous gas. Now this quotation
exactly agrees with the obfervations of officers, who
complained to me of the progrefs of contagion,
after all the care and perfeverance in the fumigating procefs. The plain reafon was, that thefe
officers had fo much faith in the practice of
fmoaking, that other precautions were very fecondary confiderations. Now when the people were
moved, large fires and a greater degree of heat
and fmoke would be diffufed through an empty
fhip; and this was thought to be certain purification. In the mean time the fever would be
dying a natural death elfewhere, among the fhip's
company; and after fome weeks, when extinguifhed of its own accord, the people would
return to their own veffel, where the fumigations
are faid to have worked wonders, though the fhip
could have no fhare in prolonging infection. It
is thus, by overlooking collateral evidence, that
error and mifreprefentation are perpetuated.

Having, in thefe remarks, given my fentiments
againft the theory of animalcules, corpufcules, or
if you chufe, living atoms, as well as the different
noxious gaffes employed to deftroy them; I will

* Papers on Infection.

go farther, left the queftion fhould be evaded by introducing another doctrine. The articles propofed for the correction of contagion, being all fubftances prone to combination, I would afk the advocates for their ufe, the following queftions :

Is there a chemical union between the gaffes, and the exhalations of difeafed bodies, as propagators of infection ?

Is contagion itfelf a fubftance, fubject to the laws of chemical attraction * ?

If one of thefe queries cannot be anfwered in the affirmative, the whole bufinefs muft be left to conjecture, or future experiment.

During the diffufion of the nitrous gas, between decks, on board the Niger, it excited *general coughing and head-ache*, which did not go off till the people had been fome time expofed to the free air. What degree of dilution do they intend to fix in order to render it refpirable ? Now if fuch unpleafant feelings were produced among people in health, is there not a degree of unpardonable levity, in carrying on this procefs, under the very cradle of a man fick of a fever, whofe

---

* The chemical Reader, will be pleafed to recollect, that thefe gaffes have not been combined with oxygene, fufficient to give them the character of acids : they do not alter the vegetable tinctures, &c.

caufe,

cause, no one dare deny, is very often a deficiency of vital air *.

Vinegar having been long in use to sprinkle about the floors of sick chambers, giving out a very grateful flavour, seems to have been the reason for our practice in sick berths in ships, to convert it rapidly into vapour, by plunging a large ball of red hot iron at once into a bucket full. A very neat apparatus for this purpose was some time

* It is curious to observe the different opinions of ingenious men, on the nature of Contagion. I now allude to a new idea on the subject, which I have just found in an article of the Appendix to the xxth vol. of the Monthly Review for August last, p. 490. The work is entitled " An Inaugural Dissertation on the Chemical and Medical History of Septon, Azote or Nitrogene, &c. by Winthorp Saltonstall, New York, 1796." The doctrine, it appears, is taken from Dr. Mitchell, professor of chemistry in the Columbia College. He denies that any of the gaseous fluids have to do with the matter of contagion, except the " combinations of *Septon* (azote) with the acidifying principle, and to manifest itself in the septous oxyd, and the *vapours of the nitric acid itself.*" Thus Dr. Smyth's preventive, the *nitrous gas*, that has been said to bear off the bays from all competitors, is here convicted of being the *guilty agent itself!!* Many able arguments are adduced in support of this explanation of contagious miasma, which bespeak an intimate acquaintance with chemistry. The Review has not quoted any means for correcting this kind of infectious gas; but I suppose that would be done by admitting factitious oxygene, or the pure atmospheric air, till the azote was saturated and changed

†   to

time ago laid before fome of the public boards, and introduced into our hofpitals and fhips. It confifts of a furnace or lanthorn, in which is a lamp ; and over it, a little earthern pipkin, in which the vinegar is contained, and by the heat of the lamp, flowly evaporated. The fmell of the vapour is very agreeable ; but I would fay no more for it. It ought to be trufted, like lavender water, on the handkerchief of a belle or a beau. The author of this apparatus, like other projectors, has his theory, which he has been frank enough to publifh. He fpeaks of the emanations

from

---

to ponderous nitric acid, when it would feek the bottom. Thus a fubject that has long evaded human refearches, will continue to give birth to new fpeculations. and is ftill involved in uncertainty.

> Afk where's the North; at York it's on the Tweed;
> In Scotland, at the Orcades; and there,
> At Nova Zembla, or the Lord knows where.
>
> POPE.

While thefe ingenious philofophers are pleafing themfelves and their readers with their arguments, on difcovering the nature of Contagion; it will be my duty to furnifh my gallant friends at Spithead with a pure atmofphere to breathe, as the beft preventive and fecurity : putting no reliance on flimfy doctrines or agents, that are held out to correct the poifon. When a more perfect plan fhall appear, I fhall be one of the firft to embrace it; in the mean time, I may be allowed to wield my little ftock of chemical knowledge, in defence of my practice; for it has been—SUCCESSFUL.

R 4

from the body being of an alkaline nature; and of courfe his vaporific plan, is to neutralize them. But vinegar itfelf, evaporated by the common heat of the atmofphere, would feem to give out vital air: in a room where this is going on, there is a manifeft change in the ftate of the atmofphere; a degree of exhilaration of fpirits is felt alfo, and if this is the cafe, it may anfwer fome of the purpofes of ventilation.

I am now very well pleafed that this difcuffion has taken place, becaufe I think our fervice muft be benefitted by it. There is fome novelty in it, and may give birth to new ideas, and future reflections, on fubjects nearly connected with the health and lives of many of our fellow creatures. It alfo fhows, that in our medical arrangements in the navy, our fyftem is incomplete. Mr. Menzies, in his report to Dr. Smyth, gives a forry account of the Union hofpital fhip : it was not only imperfectly fitted for the convenience of the fick, but ready to propagate difeafe to the healthy. It was very wrong to crowd fo many people in a malignant fever into a fmall fpace that feems to have been improperly ventilated, and badly attended on the fide of the nurfes.

If it is found neceffary in medical practice, in order to make attendants careful in their duty to the fick, to keep up their confidence in the prefervative means againft infection; I fee no

neceffity

neceffity for having recourfe to deception, or the pious fraud of a placebo. Let the nurfes of hofpitals who attend patients in infectious fevers, be impreffed with the idea, that if they fhift the fick man often, in bed-cloathes and body-linen; keep him clean in his perfon, by frequent ablution, and change the air of the ward very frequently; that it will not only recover the patient, but will infallibly prevent other perfons from being infected. Truth that foon decides doubts, will quickly affure every nurfe, that this is the only certain method of prevention; and it is the only guide that ought to regulate the conduct of a phyfician.

IN Mr. Kenning's diary, fubjoined to the artical Typhus, we have a very diftinct view of the firft appearance, the progrefs, and extinction, of Contagion, all in the fpace of feven weeks. The fhip was at fea, fo that there could be no certainty of feparation; but it was fummer, which was favourable, and it only fpread to about fifty-four cafes, including a woman, who received the infection from her hufband. About the fourth week, he remarks, that the general fymptoms were fo flight as fcarcely to confine the patient; and after the fixth week, no frefh cafes appear. The winter

feafon,

feafon, would probably have forced this ſhip to quit the fea, with a fimilar contagion on board.

It is of vaſt importahce to ſervice, to judge with deciſion, on the ſtate of infeéted ſhips, as by that means our Admiralty may know what they are to count *an effective force.* In the ſpring of the year, or in ſummer, after three weeks from the appearance of the fever, and where the early ſeparation has been duly attended to, I would not heſitate on emergency, to declare theſe ſhips in condition for fea; becauſe at this time the caſes would be ſlight, and I would expeét much good from the approaching warm weather, and probably ſome advantage would be gained by rouſing the minds and bodies of the people to aétion. In the winter feafon, however, we muſt be more cautious: the attacks are not only more frequent, but relapſes alſo; and then, when you go to fea, you have not that ſtrong nouriſhing freſh meat, which is required for convaleſcents from a low fever. If the ſhip is going to the warm latitudes, after ſeparating the bad caſes, as we did in the Vengeance, the courſe to the ſouthward will effeét what we expeét from ſummer in other ſituations: but there is a hazard of meeting foul winds before you gain the warm latitudes, and theſe with rain, might throw you back as bad as at firſt. In giving opinions of this kind, a phyſician muſt be previouſly acquainted with the diſcipline

of

of the ship, her officers, and it is fit, also, that he should know something of the surgeon ; otherwise his prognostications may be turned against his professional reputation, by his directions not having been understood, or purposely, or ignorantly disobeyed. The service also may be defeated by rash and unguarded opinions on these occasions.

MEDICINA

MEDICINA NAUTICA.

TYPHUS.

UNDER the name TYPHUS, I mean to def-
cribe, what has been called Hofpital, Jail,
Camp, Ship, Low, Slow, Nervous, Putrid, and
Petechial FEVER. Any difference that has given rife
to thefe names, in the nature of the fever, feems
to have been more owing to the peculiarities of
conftitution, fituation, climate, feafon, habits of
life, &c. of the patient, than any real difference
in the charaçter of the difeafe itfelf. We have
feen often, among a number of men living together, the fame infeçtion produce a fever, with all
the variety of fymptoms, with which authors have
defcribed fevers under thefe appellations : fuch is
our apology for the general term TYPHUS.

It is very certain that this fever is generally
fpread by contagion ; but it is equally certain,
that it frequently arifes in places where there
could be no fufpicion of communication with
infeçted perfons or cloathing. Such is the low
fever

fever that is met with in great towns, among
poor people, in low, dark, dirty, ill-aired and
damp houfes, towards the fall of the year, and
particularly in times of fcarcity, or during long
and rigorous winters.   Fevers that have arifen in
this manner, become equally infectious, as if they
had fpread originally from perfons labouring
under actual difeafe, in other fituations.   The
fame caufes we have no doubt, have at different
times produced fevers in fhips and tranfports.
There are very few of our men of war, that are
under that kind of difcipline now-a-days, that
endangers the engendering of contagion ; if there
are any, they muft be confined to fome of the
worft regulated guardfhips.   In tranfports however, under the prefent eftablifhment, and notwithftanding many improvements, we have no
doubt but contagious fevers often appear from
the caufes juft mentioned.   Thefe veffels are too
fmall, and generally crowded ; foldiers with a
number of women and children, the whole un-
accuftomed to fea.   If bad weather occurs for a
length of time, during their paffage, or even in
port, when they are obliged to live conftantly below, from falls of rain ; difeafes of this kind
foon make their appearance, and fpread by infection.

In the tenders employed to collect men from
the out-ports for the Navy, whether volunteers or
impreft,

impreft, this fever is very common. Thefe veffels being partly fitted as a prifon, for fecurity, are not the beft calculated for health. At the beginning of a war, the idle and profligate are either im-preffed, or come to the regulating officer to enter; all defcriptions are therefore mixed together in the fame deck. Many of them bring, from their hiding places, the feeds of contagious difeafes in their cloathes, which are fpeedily extended to others. No method has yet been practifed to cloathe thefe people; and they often fleep for weeks upon the boards without a bed : this, joined with the crowded ftate of the tender, never fails to render them fickly, if the paffage happens to be long, and the weather bad.

Receiving fhips, on their prefent eftablifhment, are alfo very liable to fofter the powers of contagion. We trace infectious fevers more frequently to them, than to all other fources ; witnefs the account given by fome of the Orion's people. Befides the new-raifed men fent from tenders, all recovered men from hofpitals are fent hither till their fhips return to port, or till they are otherwife difpofed of. No regular difcipline is likely to be preferved for the fecurity of health, among people of fuch different orders and defcriptions; the feeds of contagion are therefore fpread among them, before they are detected.

I have

I have already narrated the firſt train of ſymptoms that indicate infection; but they vary in
different conſtitutions, both in their manner and
time of appearance. Slight ſymptoms of indiſpoſition, ſcarcely noticed by either the patient or
his friends, are ſometimes the harbingers of this
fever. A langour, and inaptitude to motion,
rather than pain of the extremities; being more
ſenſible than uſual to external cold; heavineſs of
the eyes and forehead, rather than head-ach;
yawning, dejection of ſpirits; ſleep diſturbed, and
not refreſhing; appetite impaired; and diſagree-
able taſte and ſmell; are among the moſt early
ſymptoms of typhus.

At other times, the attack is more rapid and
violent; with great depreſſion of ſpirits, even to
fainting; vertigo, palpitation of heart, convulſions and delirium. The hyſteric affections,
or *globus*, as it is called, and fits of epilepſy, are
frequent; but I have not obſerved that they prog-
noſticated a more dangerous diſeaſe.

The *globus hyſtericus*, was a very common ſymptom, among the Iriſh landmen, in the Vengeance, and alarmed ſome of them with the dread
of inſtant ſuffocation.

The ſenſation of cold, on the ſurface of the
body, in ſome caſes continues for a length of time;
but in others wears off in a few hours, and is
alternated with heats, that ſeldom produce
moiſture

moifture on the fkin, and fweating. The fkin itfelf foon acquires a fallow hue, and is dry and fhrivelled. The fenfation of cold is not however a conftant attendant of the acceffion; I have feen many, where no rigour was perceptible.

When the feeling of coldnefs on the fkin is in a greater degree than ufual, the fhivering and fhaking of the body feem to be in proportion; the colour difappears from the countenance, and the cheeks and lips are pale or livid. Thefe fuc-cuffions of the body indicate the vaft lofs of muf-cular power: every mufcle is more or lefs agitated by this particular kind of tremour, as if each fibre that compofed them acted for itfelf, and no longer fubfervient to voluntary motion.

The appearance of the tongue is various; fometimes it is moift, and other times parched and foul, with, or without thirft. The breath is fre-quently fo difagreeable, that the patient conceits the bad fmell to rife from fomething elfe: the dif-charge of mucus from the noftrils is now and then totally ftopped. A brackifh tafte is felt in the mouth; or what has been diftinguifhed by the tafte of copper: there is a conftant naufea and uneafinefs at ftomach, that approaches to vo-miting. With thefe fymptoms, there is an op-preffion about the breaft, with a difficult refpi-ration, anxiety, and fighing.

Acute

Acute pains refembling rheumatifm, are felt in the limbs, loins, fhoulders, and breaft.

The eyes are dull, or inflamed, the pupil dilated, fuffufed with tears; and the whole countenance exhibits the look of affliction in both body and mind.

'The pulfe, at the beginning of this fever, in the fpace of a few hours, will occafionally vary confiderably. It is weak, quick, tremulous, and intermitting; but at other intervals, more ftrong, full, equal and regular. The mufcular action of the arterial fyftem partakes of the general debility; indicated by this irregular and variable pulfation at the wrift, palpitation of the heart, and throbing of the temporal arteries.

When thofe rigours or chills, alternated with heat, and fuccuffions of the body which ufher in the fever, have difappeared and fubfided, the patient is left in a ftate, more or lefs approaching to ftupor; and from which the difeafe has derived the name of *typhus*.

Some anomalous fymptoms are occafional attendants of this fever; fuch are, catarrhal complaints, and fome others indicating pulmonic inflammation, The difficulty of breathing has been fo confiderable, and tightnefs acrofs the breaft, with acute pain, that venefection has been unguardedly practifed, and repeated even to the

S   ' third

third time, but with manifeft difadvantage. I
have frequently been taken to the bed-fide of the
patient, and informed that he was labouring under
peripneumony, when a very few queftions, and
the appearance of the fick man, foon gave me a
different opinion of the difeafe.

The fever with which the London was infected,
had many fymptoms refembling thofe of pulmonic
inflammation; and it certainly, in a number of
cafes, was taken for pleurify. Catarrhal complaints are alfo liable to deceive us: the effects of
contagion are often taken for a common cold at
firft, becaufe the indifpofition fucceeds to the
expofure to weather, by which means the opportunity is loft, of putting a ftop to the difeafe
in its earlieft ftage. But it is only in the hands
of inexperienced, ignorant, or inattentive ob-
fervers, that a fever of a contagious nature can
flide on without being detected. For the fame
reafons, I have feen patients with petechiæ upon
them, fent to Haflar with the ticket marked,
meafles: a cafe or two, from the fame error, were
alfo fent on board the Charon hofpital fhip.

In fome bad cafes of the typhus, which occurred among our people, after communication
with the French prifoners, there feemed to be a
greater quantity of the biliary fecretion, than
what was ufually met with, in fevers which appeared in the winter feafon. The weather was extremely

tremely hot at that time; and probably the
fyftem of the *vena portarum*, partook of that peculiarity which affects it in warm feafons and
tropical countries. But this not being always
the cafe, I can give no reafon for the appearance
of jaundice, under other circumftances of the
weather: I have often feen the eyes and fkin, with
the urine, as deeply tinged by the bile, in typhus,
as I have remarked it in the yellow fever of. the
Weft Indies.

The fever among the worft cafes which I examined in the Portugueze fhips, was diftinguifhed
more by the tremulous motion of the tongue, *fubfultus tendinum*, *dyfphagia*, and *fingultus*, .than is
ufually met with among our feamen.

The fymptoms which I have fo far defcribed,
and which generally appear in the firft three or
four days, I would call *the firft ftage of Typhus:*
becaufe this diftinction will lead to very important regulations in the practice.

The *fecond ftage* of this fever, or what I would
call *the ftage of Stupor*, is that ftate in which the
patient is left, after the rigours, &c. have totally
fubfided, without any particular pain of any part.

It is marked by almoft univerfal mufcular debility, and little defire of exerting it. There feems
at times, fuch a.want of attention to fenfative motions, or external impreffions, that the patient
would perifh, from inanition, were he not urged

S 2 to

to take nourifhment. He commonly lies on his back; ftarts up when fpoke to, as if furprized, and requires fome time to recolleét himfelf before he can give diftinét anfwers to what has been afked. The fleep is difturbed with dreams, and not unfrequently, fuddenly terminated, when the countenance exhibits ftrong figns of horror and agitation. Muttering is alfo common in fleep; and the appearance of being bufily employed in particular affairs. In people endued with greater fenfibility of the nervous fyftem, figns of grief are expreffed, by weeping, fobbing, palpitation, and fometimes convulfions. The tafte is altogether obliterated, and the fur on the tongue becomes brown and black: the tongue is alfo tremulous, often dry; eruptions, turning into little fcabs, appear about the lips; they are frequently included among figns of convalefcence.

During this extreme degree of debility, the frequency of the pulfe is fometimes fo confiderable, as not even to be counted.

The fpots on the fkin, petechiæ and vibices, which fo frequently attend typhus, appear at different periods of the difeafe. I have never confidered them of fo much importance as fome writers have done: they occur occafionally in very mild cafes, and are wanting in others of the greateft danger.

Sir

Sir John Pringle, in his Difeafes of the Army, mentions an unpleafant fenfation on the end of his finger, which continued for fome minutes after feeling the pulfe of the patient. He was at firft inclined to refer this to the force of imagination, but found that others made the fame remark. Some authors have repeated this feeling in their account of the fymptoms, but whether only copying, or confirmed by their own obfervation, does not clearly appear. For my own part, nothing of the kind ever occurred to me. Surely nurfes or attendants, in moving and lifting the fick, would fometimes take notice of this circumftance: it, however, has never been told from their evidence. It appears to me to have originated from affociating the dread of infeftion with the touch of the difeafed fubjeft, and is altogether fanciful. When vifiting fever-patients, whether in the hofpital-wards, or in the fick-births of fhips, it has been my cuftom to feel the pulfe of many, that I did not expeft to learn any thing from, and to pafs others where I knew it to be of no confequence. What I have moft cautioufly avoided, is breathing immediately over the patient. But even that precaution will be often in vain: I have met with many fick, who ftart up in bed, as foon as I approached, feemingly in furprife, and throw the bed-cloathes from them, or put their mouths clofe to mine. I cannot however forbear mentioning

S 3　　　　　a praftice

a practice with some medical attendants, whether physicians or others, of keeping at a distance from the patient while they examine him, or ask him questions. But a typhus patient is sometimes so low, that to hear him speak distinctly, it is necessary to lay the ear to his mouth. If there are physicians who think otherwise, I envy them not their peace of mind. But a medical attendant, who approaches the sick-bed of his patient, in this manner a dupe to his fears, and the slave of his apprehensions, appears more like a recorder of Newgate, about to pass sentence of death on a criminal, than a man who bears the commission of Providence, and the stamp of sympathy, to administer the duties of his profession, for the comfort and relief of a fellow-creature.

I will now describe some of those symptoms which more especially indicate danger. The first is early delirium, the pupil of the eye dilated and insensible: delirium may be reckoned only an increased degree or variety of stupor; the more early it appears, it shows a more aggravated disease. It has been observed so soon as the first attack, or in a few hours, and death has followed on the second or third day. It has been sometimes so sudden, as to be taken for apoplexy, by attendants not much experienced to the effects of contagion, where a number of men are exposed at once to its influence.

Hiccups,

Hiccups, and fubfultus tendinum; the fœces and urine, &c. paffing off infenfibly, are figns of danger; but I have feen patients recover from them all. But the more early they occur, they are the more to be dreaded: an irregular weak pulfe, under thefe circumftances, is always dangerous.

There are alfo conditions of the excretions of the body, which point out a fevere difeafe. The urine that was before pale, becoming black and grumous, is a bad fymptom: fo alfo I confider hæmorrhages, a yellow-darker colour of the fkin, with increafed fœtor of the breath, &c. thefe conftitute what has been called putrefcency.

Suppreffion of urine, and difficult deglutition, are parts of the extreme mufcular debility. Deafnefs has not been confounded with a ftate of danger.

The favourable fymptoms are, the gradual recovery of fenfation and perception; often at firft, known by the patient longing or calling for fome particular kind of food; the fleep lefs difturbed; the tongue becoming moift and of a livelier red on the edges, and the pulfe more equal and lefs frequent. But thefe appearances return by fuch flow degrees, that a nice difcernment is required to turn them at once into favourable omens.

TREATMENT.

S 4

TREATMENT.

The treatment of Typhus in the early ftage, or during the firft figns of infection, has been already mentioned. This is the period when moft is to be expected from our art ; if we fail here, in fpite of all our efforts, the difeafe feems to run a determined courfe, and is protracted to an uncertain duration.

Whatever has been pointed out to us, as indications for the cure of fever, by ingenious phyficians, appears to me to have been built, more on hypothefis and inconclufive reafoning, than real matter of fact. After an attendance on fome thoufands of cafes, in all circumftances and fituations, diverfified by fex, age, conftitution, feafon, and climate, I have no doctrine to offer on the fubject, that may not be liable to moft of the objections urged againft preceding theories. We have, hitherto, no certain method of cure ; the nice and intricate phænomena of fever, fhew that the tafk is arduous, if not prefumptuous.

A febrile ftate, as it is defined, after a fenfation of coldnefs, frequent pulfe, increafed heat, injury of fome of the functions, efpecially with fome diminution of voluntary motion, is found to take place in very different conditions of body. Whether, therefore, the fubfequent fever be of a low,

or

or inflammatory nature, is attended with fome increafed excretion, or an eruption on the. fkin, is
intermittent or continued in its type; ftill the
phænomena which ufher in the difeafe, are more
or lefs the fame. It is true, that fevers attended
with inflammation of particular parts, have a fuller
and harder pulfe, than we find in typhus; with
alfo the difference of high-coloured urine, while in
typhus it is little changed; yet in both difeafes
they would only feem to be fympathetic, with a
peculiar difpofition of body. They are indeed fo
much alike, as liable to be converted into one
another; hence Dr. Cullen has given the genus
Synochus, " intio fynocha, progreffu, et verfus
finem, Typhus." The fever which precedes the.
eruption of the confluent. fmall-pox, is exactly
of this kind; and at the acceffion, there is no
fixed criterion, by which we can tell whether the
future difeafe will be diftinct or confluent. In the
hectic of phthifis, we fee it regularly introduced
with more or lefs chill, fucceeded by heat: the
fame phænomena take place during the formation
of pus, and in fhort, whenever there is condition
of body attended with fever.

In all thefe different genera of fever, we therefore obferve fomething peculiar to the whole, and
the fact is univerfally admitted. What then are
we to fay produces the phænomena of fever, or
what their proximate caufe?

It

It is no unfafhionable way of getting quit of
an intricate point in the doctrine of fever, to refer it to the laws of the animal œconomy, as an
ultimate fact.   We have feen thefe phænomena
of fever, occur in very different ftates of the
body, the forerunners of an inflammatory difeafe,
as well as one of diminifhed excitement. Would
it not, therefore, be right to conclude that thefe
fymptoms are nothing elfe but affections, fympathetic of certain offending powers applied to the
fyftem ; congenial with it, acting in unifon to its
feelings, and as naturally excited by them, as
tickling the pharynx excites coughing : or irritating the eyes occafions the fhedding of tears.

Thefe febrile actions are primarily belonging to
the nervous fyftem, and by confent communicated to other parts.   The patient often dies in the
cold fit of an ague ; delirium is frequently among
the firft fymptoms of infection, and almoft inftant
death has been the confequence.   Now what we
mean by a cold fit, is nothing more than the diminifhed energy of the brain : ftupor and delirium are alfo owing to this diminifhed energy,
but in a greater degree.   The hot ftage, is the
confequence of the preceding cold one : the debility and fenfation of cold continue, till the fenforial power or fpirit of animation is fufficiently
accumulated, to excite the fanguiferous fyftem
and other parts to an increafed action, by which

3   means

means the heat is reftored on the furface of the body, and the flow of fweat takes place. But if the powers of life are fo far depreffed, and the fpirit of animation can be no longer recruited and accumulated, then the patient expires in the cold. fit : or, by a quantity of fenforial power, ftill fit to preferve life, but not equal to reftore the energy of all the functions, he remains in that ftate of ftupor, which conftitutes the fecond ftage of Typhus.

We have divided Typhus Fever, into two ftages : the firft comprehends the early fymptoms, or as long as chills, and heats alternating with each other, can be diftinguifhed : the fecond ftage, is that ftate of ftupor that follows; but fometimes appears alfo, without any previous chills and heats having been obferved.

I confider this diftinction of real practical utility, for it is only during the continuance of the firft ftage, that we fuppofe the difeafe to be capable of cure ; we call it *treatment* what applies to the fecond ftage.

The remote caufes of fever, are certain hurtful powers, liable to affect the human body and derange health ; which, from laws peculiar to animal life, excite the phænomena of fever, and which in themfelves are only fymptoms of the difeafed ftate. This difeafed ftate confifts of impreffions received from the remote caufes, which will

will remain till they are expelled by others, on which the method of treatment depends. This comes near to the doctrine of one irritation overcoming another : the early fymptoms of Typhus are fuppofed to be eafily moved by fuitable remedies ; and are a proof that they make no very lafting impreffion on the nervous fyftem : but thofe fymptoms that conftitute the ftage of ftupor, fhow. that the ufual habits, motions, and appetites, are overcome or forgot, and nothing but a length of time, days, and even weeks, are required to reftore them.   We fupport and nourifh the body during this inexplicable condition in the beft manner we can ; and we adminifter ftimulants to preferve and excite its energy, which forms the treatment ; but we are ignorant of any medicine that can renew the healthful exercife of the functions, at any period during this ftage of the difeafe.

The cure of an ague by the bark, or other remedies, is to fupport the healthy actions for a length of time, at leaft till the period of the cold fit is over ; to renew this fupport for fome time, at the proper hour of the expected acceffion of the paroxyfm ; and to continue this fupport till all the actions, motions, and habits of the fyftem, have acquired their accuftomed vigour, which conftitutes health.   The cure of the firft ftage of Typhus, is analogous to this practice ; and whatever has been found ufeful in the fecond ftage,

will

will likewife anfwer to this explanation. We are
of opinion, that the whole act by the *production*
of fenforial power.

The fuccefs of an emetic on the firft figns of
infection, has been acknowledged by moft writers
on this fubject. The effect, however, has been
differently explained. It has generally been accounted for, by evacuating the contents of the
ftomach, and freeing that organ from any morbid
irritation. It has been thought by fome great
authorities, to act by expelling the contagious
matter from the ftomach, which, they affirm, there
generates the fever. The good effects which we
have experienced from emetics, have been equal
to thofe recorded by other phyficians ; but we
fuppofed at the time, that under certain circumftances, they were fingularly fuccefsful. I fhall
narrate them. The vomiting ought to be free ;
fo as certainly to evacuate the whole contents of
the ftomach, without however, endangering any
pain of that vifcus. It is eafily excited by a grain,
or a grain and a half of tartarized antimony, joined to fifteen or twenty grains of the powder of
ipecacuanha. It is preferable to give the vomit
in the evening, becaufe, by fleep following quickly to the operation of the medicine, unpleafing
fenfations are avoided ; and it protracts the recurrence of the rigours. When the ftomach feels
perfectly fecure againft farther retching, a gentle
dofe of the pulvis antimonials combined with
opium ;

opium; or a draught made with vin. antimon. to
fixty or feventy drops, with thirty of the tinct.
opii, ought to be taken at bed-time, and followed by a few glaffes of generous wine, wine-whey,
or negus, with a bit of bifcuit or toafted bread.
If thefe are difagreeable to the fick, fome other
more grateful food may be found, but it is neceffary that the ftomach fhould not be left empty,
after the vomiting. The heat of the chamber,
or ward, ought to be equal to our fummer heat.
Next morning, our patient ought to make an
early breakfaft, on what he is accuftomed to eat :
if free from feelings of coldnefs on the furface,
or other febrile languors, he may ufe the cinchona, in either tincture, powder, decoction, or
infufion, with a glafs of wine at intervals. If,
however, flight rigours return, with want of appetite, he had better go on with the antimonial
medicine, under the fame directions as given with
the treatment after infection. The dinner ought
to be at the ufual hour; and if he is in the habit
of drinking wine, in the fpace of two hours he
may be allowed to finifh his pint. The food
fhould be light and eafy of digeftion; at the fame
time, it muft be of that kind that the appetite
will prefer. Although the defire for food might
be fufficiently good, yet in meat as in wine, the
indulgence is meant to fupport the body effectually, without endangering that langour and
debility,

debility, which are the certain followers of excefs in either. The antimonial and opiate, are to be continued at bed-time ; and the bark may be added next day to this method of cure. The ftate of the bowels is to be attended to.

In warm weather the patient may amufe him-felf out of doors, in any exercife or employment he pleafes ; but external cold ought to be cau-tioufly fhunned during this procefs, and the ufe of antimonial medicines. The mind ought, if poffible, to be fo engaged, as never to be allowed to brood over its own feelings, or to re-call any unpleafant affociation of ideas. The remem-brance of a friend or relation in the fick-bed, from whence infection was communicated, is apt to renew former impreffions ; the rigours return, and ficknefs and naufea follow.

In the treatment of the early ftage of fever, or even at the firft figns of infection, Dr. Lind every where lays great ftrefs on the ufe of blifters. He fays, " when a patient came into the hofpital in the evening, and had a blifter applied, he was always better next morning :" this continues to be the practice at Haflar indifcriminately. I believe, fince Dr. Lind's time, no writer on the fubject had added his teftimony, in fuch general terms, in favour of bliftering. Much of the fuccefs which he attributed, I apprehend, may very eafily be refolved by another reafon. When a failor in

fever

fever is fent on fhore from a fhip to an hofpital, from a gloomy fick-birth on board, into the clean and well-aired bed of a fpecious ward, with an attentive woman to nurfe him; if any fenfe of his fituation remains, I think the tranfition he undergoes, could not fail to make him better: if he has no fenfes left, it is almoft enough to reftore them. But I can very well fuppofe, that blifters may be ufeful : the abforption of a confiderable quantity of cantharides from a bliftered fpot, may, as a general ftimulus, affect the whole body; and it may be fo permanent too, as to excite the fyftem a fufficient time to refift the difeafed actions : my own obfervations, however, do not juftify this conclufion.

Dr. Lind's practice of bliftering in fevers, re-mained in its full force when I joined Haflar Hofpital ; but the cuftom was too indifcriminate to be deemed rational, and it was continued, out of refpect to his authority, by the whole medical gentlemen of that eftablifhment. The once cele-brated James's powder was adminiftered at the fame time, but lately changed for the pulv. antimonialis of the London Pharmacopeia. Lind was fpare in the ufe of his cordials; probably he was afraid of trufting the exhibition of wine to the nurfes, who are intrufted otherwife with the giving of medicines; and what I confider an un-pardonable defect in the inftitution.

In

In local pains, fuch as thofe of the breaft and fide, we have feen blifters followed with fpeedy relief.

After the operation of the emetic, we have often joined ammonia with the opiate: in cafes where there feemed more than ufual torpor, and depreffion of fpirits, I have thought it more effectual than any preparation of antimony.

Of a febrifuge power refiding in antimony, our obfervations and experience do not enable us to fpeak in confirmation; nor do we fuppofe, that fuch a power is to be found in any other medicine whatever. The quality it has been faid to poffefs, of determining the blood to the furface of the body, and thereby removing the conftriction of excretories, and foftening and relaxing the fkin, are only proofs of its exerting a ftimulant power. During its ufe the patient fhould either be kept warm, or lie in bed, and drink moderately of diluting liquors. This method renders it lefs liable to act on the ftomach and bowels. It is only in the firft ftage of Typhus, that this medicine appears to be ufeful: we have no doubt, when taken as has been directed, it has had very confiderable effects in our practice. After having produced full vomiting in the firft inftance, I think the milder preparations only fhould be continued. When the tartarized antimony has been well rubbed with magnefia or chalk, which I

T   fufpect,

suspect, abstract a part of the acid, while it is exhibited, their powers together are much like the effects of James's powder, or the antimonial powder of the dispensary. Such has been our practice in counteracting the early symptoms of infection, and in curing the first stage of Typhus.

The second train of symptoms, are those which attend the stupor, and are chiefly distinguished by the loss of muscular power, more or less dilirium, and other signs of debility. Hence, lying on the back; the loss of voice; the tremulous motion of the tongue on putting it out; indistinct vision; want of perception; muttering to himself; catching and picking the bed-cloaths; moaning, sighing, weeping, sometimes laughing; difficult deglutition; paralysis of the spincter muscles; tremors; convulsions; syncope on being raised upright: singultus; cold sweats, as they are called; grangrene of blistered parts, of wounds, and sometimes of the extremities; hæmorrhages, intolerable stench of the breath and excretions: the skin dusky or yellow, petechiæ, vibices; vomitting of a black matter; the stomach rejecting every thing immediately as swallowed; profuse diarrhœa; dark coloured urine; glassy appearance of the eye; the eye-balls fixed, sunk; the countenance shrunk, lengthened, ghastly, and discoloured; the inside of the mouth, and tongue, black and parched; the mouth drawn aside, and

tongue

tongue hanging out; deafnefs, or hearing very
acute; the jaw fallen; immobility of the joints;
watchfulnefs, or conftant fleep; heaving of the
breaft; rattling of the throat; pulfe felt only at
intervals; *conclamatum eft.*

The treatment of this ftage of Typhus has been
almoft confined to ftimulants. In books on the
fubject, however, we do not meet with much
fatisfaction in the felection : little difcernment is
fometimes to be found in giving to one article a
preference over another; and the exhibition of the
whole feems to have been too often left to indi-
cifion, or a random practice. Phyficians have
particularly fallen into this error, from vifiting
their patients at too long intervals, by which
means they have been unable to judge of the
effect of their medicines, and the report of the
whole has been unfortunately left to ignorant attendants, or nurfes.

The only indication in this ftage of the fever
appears to be, to fupport the body by the nourifhing and ftimulating articles of diet and medicine, that are moft grateful to the fick, that fit
eafy on the ftomach, that procure refrefhing fleep,
recruit the fpirits, and that do not exhauft by any
indirectly debilitating effect.

We fhall firft mention the articles which belong
to diet. It is not eafy, at all times, to fuit the
defires and unaccountable longings of the fick;

but, if poffible, they ought to be indulged; even
to the moft out-of-the way articles, if I may ufe
the expreffion.   While they can choofe for themfelves, it is prefaging a favourable iffue. A young
man, about eighteen years of age, after fixteen or
feventeen days confinement in bed, and juft beginning to fhew figns of convalefcence, difcovered
the firft return of recollection and appetite, by
calling for a beef-fteak from a beat bull.   A
common beef-fteak was immediately got ready;
but he was aware; from its being got fo foon,
that it could not be what he wanted.   The
carrier, however, that went to a market-town, at
the diftance of fifteen miles, was ordered to bring
one when he returned.   Of this the patient eat
heartily, and relifhed it as much as if he had got
his defire.   His appetite for other kinds of food
was progreffive, and he recovered apace.

The diet ought to be always nourifhing; and
as folid animal food is feldom wifhed for, before
convalefcence, broth made from beef or mutton,
with the fat fkimmed off, and a little acidulated,
with the juice of lemons, or even port wine, or
claret, is, of all fpecies of cooking, the moft
eligible.   To this broth, may occafionally be
added, fome grateful aromatic, fuch as mace,
cinnamon, nutmeg, ginger, or even common pep-
per.   Analogous to this broth, are all animal
jellies, but weaker nourifhment, fuch as that of
                                                      calves

calves feet.   It may be well imitated at fea, when·
there is no frefh meat on board, by the common
portable cakes of foup.   Thefe cakes of foup
ought to be diffolved in the boiling water as foon
as it is moved from the fire, but they ought
never to be boiled with it, which renders the
whole unpleafant, and gives it the tafte of glue.

The flightly fermented juice of oatmeal, called
fowens, to which wine, and a little fpice, are
added, we have often feen acceptable in a fick-
birth; and it ought to be a conftant article in
fick diet, at fea.

When milk can be procured, muftard whey
may be ufed as a change from other articles.   It
is, when properly made, a very grateful ftimulant;
and is one of the beft fubftitutes which I am
acquainted with for purgative medicines, as it
will generally keep the bowels open, without
feeming to do fo, fo gentle is the operation.

When light puddings, cuftards, &c. are relifhed, they may be fafely indulged in.   They are
nourifhing, and eafy of digeftion, without any
hurtful quality.

The thirft is beft quenched by fruit, fuch as
oranges, apples, goofeberries, currants, rafps,
prunes, melons, &c.   They have the advantage
of being pleafant, and fharp to the tafte; and, if
they do not diforder the bowels, which they are
apt to render too lax, they may be ufed *ad*

libitum. When thefe cannot be procured for drink, diluted wine, negus, porter, bottled alc and fmall beer, cyder, perry, &c. are generally very acceptable in this fever.

Having enumerated thefe articles in the diet- etic method of treatment, I have to regret, that too few of them are to be met with in our hof- pitals.   In one, which I have had fome acquaint- ance with, it has been told to me, that to fever patients, not lefs than twenty guineas were ex- pended *per diem*, for Dr. James's powders, upon an average, for a whole long war; yet, fuch a delicacy as a piece of roafted meat, a pudding, a jelly, or a cuftard, has never been feen within thofe walls !   Thofe phyficians, who can cure difeafes which fpring from mifery and want, with- out articles of nourifhment and fupport, muft have an unbounded confidence in the wonder- working powers of medicine.   The relief, how- ever, which the braveft of the human race, the Seamen of Great Britain, have obtained, by a contrary treatment, has made me a fceptic on the fubject.

This argument may be extended to fome cha- ritable inftitutions in great towns, where putrid fevers are frequent.   The account of the poor family, as given under the head Contagion, is a melancholy proof.   I never faw a corporation- dinner, but thefe difeafed and miferable creatures

came

came into my mind. When leaving Newcastle, at Christmas 1792, I addressed the Governors of the Dispensary, on the subject of a dietetic department; and was informed by some of them, that it should meet with their warmest support: the fate of this address has not come to my knowledge. This Institution took its rise from the benevolent exertions of Dr. John Clark, and Mr. John Anderson the surgeon; the former is the celebrated medical author. The patients of this charity are now visited by all the physicians, with the most punctual attendance; they are, moreover, men high in professional character. In point of subscription, it is nobly supported; the town itself abounds with affluence, and is noted for charity and the Christian virtues. If it ever should be my fortune to return, when I have discharged my duty to my Sovereign, my Country, and her Defenders, I shall not fail to revive the subject.—But to our purpose:

The body ought to be washed all over every morning, with water about 50° of Farenheit. This may be done with a sponge; but I prefer a clean towel, which is to be dipped in the water, and rubbed over a piece of soap, to make it more effectual in purifying the skin from any perspirable matter. After this operation has been attentively performed every morning, the body and bed linen ought to be shifted; taking care, if the

patient

patient is very weak and reduced, to move him
gently, and to keep him, as much as poffible, in
nearly an horizontal pofture, to prevent fyncope.
I have known a patient expire immediately by
much fatigue from unneceffary motion, and being
kept long erect by the officious intermeddling of
ignorant attendants, or obtrufive vifitors. If the
ablution, or wafhing of the body, and the fhifting, is rightly performed, for it conftitutes one of
the niceft duties of fick-nurfing, the patient will
feel himfelf enlivened and recruited; fometimes
difpofed to take nourifhment, which is frequently
followed by refrefhing fleep.

The medical attendant at every vifit fhould be
very careful in examining and watching the duty
of the nurfes, and others about the fick; with
regard to the ftate of the inteftines, and the
urinary difcharge. I remember among fome cafes
of my early practice in Typhus, under the direction of fome eminent phyficians; when from
neglect in attending to the evacuation of the
vefica urinaria, death unexpectedly happened.
This ought to be particularly guarded againft,
after the application of blifters; and, if medicine
fails, recourfe muft be had to the catheter, every
fix hours, or oftener if the patient takes much
liquid aliment. This painful fymptom feems to
be owing, either to the directly ftimulating power
of the cantharides, or to the torpid or weak ftate
of

of the fphincter, or mufcular fibres of the bladder, after the preceding exceffive excitement from
the fame caufe.   The finer powder, of thefe in-
fects is taken into the circulation, feparated by
the kidneys, and thus carried to the neck of the
bladder by the urine.   The moft effectual remedy
in our practice is the mift. camphoræ, well diluted, and its beft affiftant is æther in muftard
whey.   Camphor is a refinous fubftance, though
with fome portion of gum, not eafily fubdued by
our fluids: and I conceive it acts by being carried unchanged to the neck of the bladder, where
it exerts a ftimulant power, by exciting the action of the detrufor urinæ, and thus expelling its
contents.   Acrid glyfters are alfo ufeful, and the
turpentines; the warm bath; or plunging the
patient fuddenly into cold water, into which nitre
or common falt has been juft thrown, to reduce
the temperature, by their folution.   This acts by
the affociation of the furface of the body with
the mufcular fibres of the vifica urinaria.

A regular alvine difcharge is certainly of great
importance in this fever.   In the exhibition of
medicines, or other articles for that purpofe, we
ought, however, to be very cautious, as their debilitating efforts are to be dreaded and avoided.
The ftools are commonly offenfive, and there can
be no doubt of the neceffity for preventing them
to accumulate.   Glyfters afford but a partial
evacua-

evacuation: I think laxatives ought to be given by the mouth, if nothing contra indicates this method. They tend, in their paſſage, to excite the biliary and pancreatic ducts, and mildly ſtimulate other excretories in the inteſtinal canals. We have ſeen nothing preferable to the tartarized infuſion of ſenna, made pleaſant by a couple or more tea ſpoonfuls of brandy, or even aromatic tincture of any kind. The medicines for this purpoſe ought, however, to be varied to the ſtate of the ſtomach, and made agreeable to the patient. Their exhibition muſt be regulated by the late preceding evacuation, and not to wait till a longer period of dejection urges their uſe. By carefully attending to this circumſtance, laxatives can be given in ſmall doſes, with ſome certainty in their degree of operation; and by which means the danger from their weakening effect will be very much obviated. We are of opinion, that an eaſy motion ought to be procured every twenty-four hours; and never to extend beyond thirty-ſix.

When a diarrhœa ſupervenes the uſe of laxatives, or purges, or any particular kind of diet, attention ought to be directed to the particular articles which may have induced it. Thus, an exceſs of the ſummer fruits will often have this effect; and a diarrhœa, ſometimes obſtinate to be cured, frequently follows purgative medicines,

Wine,

Wine, abounding with tartar, has alfo this effect.
When it arifes from an excefs of acid, it may be
difcovered by the yellow colour of the ftools,
from its action on the bile. The wine, or other
fufpicious liquor, ought therefore to be changed,
the fruit muft be given up, and the acidity muft
be corrected by the chalk mixture, magnefia, bit-
ters, and opium. If opium comes to be rejected,
when taken by the mouth, it muft be given in
glyfters, and in large dofes at once, that the
diarrhœa may be effectually checked at the firft
outfet. In the remote ftage of Typhus this complaint ought to be narrowly watched: the debility from it, is fometimes inconceivably rapid, and
the patient reduced irrecoverably.

I fhall now give my opinion, and the refult of
my experience, concerning bark, wine, opium,
camphor, ammonia, æther, blifters, and fome
other articles employed in the treatment of the
laft ftage of Typhus.

I have already remarked, that in our opinion,
no article of medicine, which has come under our
adminiftration, poffeffes the power of reftoring
the ufe of the functions, under any form of prefcription, that we are acquainted with. Much
has been written and fpoke of the efficacy of the
cinchona or Peruvian bark, in curing putrid and
peftilential fevers. Authors have, however,
thrown out their ideas on the fubject fo vaguely,
that

that at this moment we are left to guefs, at what
precife period of the difeafe, and under what
preffure of circumftances, it can be moft fuccefsfully adminiftered. It is true, that fome phyficians have told us of their giving it in the early
part of the fever; but they have not added, in
the fequel, that it cut the difeafe fhort, or prevented the ftage of torpor or ftupor; or that this
ftage was cured, either directly, or after a fmall
fpace from its exhibition. They have neglected
to inform us, that they watched the ftate of the
pulfe, at fhort intervals; or that a return of appetite, the recovery of perception, or fleep, found
and refrefhing, were the conftant or general attendants of this mode of practice. Yet there was
fomething very favourable to this method of cure,
and the reputation of the medicine, among many
of their cafes; for we hear them fpeaking, in a
tone of exultation, at the large quantity of bark
which they had been able to throw into the ftomach in the fpace of twenty-four hours; often to
the amount of two ounces for days together. Yet
in none of the cafes on record, when this practice
has been carried to its fulleft extent, have we
any fair detail of thofe fymptoms which firft indicate the certain, though flow recovery of Typhus patients. Among the greater number of
bad cafes which have come under my own care,
I have generally found the ftomach fo weak and
irritable

irritable from the beginning, that to have perfifted in the exhibition of the cinchona, would
have been little better than putting my patient
to the rack. The fever, in their hands, feems,
therefore, to have run on to an uncertain length.
But we have feen, in numberlefs patients, where
bark had been exhibited in large dofes for a few
days, till at laft the ftomach would be fo op-
preffed, that the whole has been difgorged : and,
upon examination, not a dram of the fubftance
could be fuppofed to have undergone the flighteft
change from the digeftive procefs. At a time of
the, fever, indeed, when the defire for food is
almoft extinguifhed, and when the moft delicate
is taken with indifference ; does it appear like
reafoning with a knowledge of the pathology of
the human body, to fuppofe, that the grofs powder of the cinchona, abounding with refin, can be
a fit fubject, at that period, for the digeftive·
organs? Hence we account for its being thrown
up after, unaltered,· although it had remained for
fome days in the ftomach. It is not, therefore,
a juft calculation, to fay that the bark cured
every cafe where it was given, becaufe the patient
recovered. Yet the average has generally been
given in this manner ; and the exception applies
to other medicines that may have been popular
as well as this long extolled febrifuge.

In all fituations and feafons, where I have feen
the contagion of Typhus extend itfelf among a
number

number of perfons, the firſt cafes have always
been the moſt fatal and dangerous. Now, if we
were to compare any method of cure with another,
it would be proper to ſelect an equal number,
taken ill at the ſame period of infection; even a
week's diſtance, will often exhibit a ſurprizing
difference in the nature of the ſymptoms. Among
a ſhip's company, a regiment, or any other body
of peeple living together in a ſimilar way; we
have uniformly obſerved, that the attacks of the
diſeaſe become gradually milder, till at laſt it
ſeems incapable of communicating infection. This
point having been clearly aſcertained from long
experience, enabled me to give my official opinions
on the progreſs and extinction of contagion, with
a degree of preciſion, that could only be acquired by a perfect acquaintance with the ſubject.
Now phyſicians, in recording the ſtate of health
in fleets and armies, ſhould be mindful of all
collateral circumſtances; leſt, what they may relate as a ſuccefsful method of cure, ſhould fail in
the hands of others, who may try it, at a different
ſtage of infection, and when more virulent ſymptoms ſhewed a more dangerous diſeaſe.—But to
return to our ſubject.

In mild cafes of Typhus, where the diſeaſe never
arrives at the ſtage of ſtupor; or where there are
regular remiſſions; I think bark is uſeful, and
favours recovery. It has therefore, for a long
time, been our practice to confine the uſe of cin-

4

chona to the mild degree of the difeafe, and during
the convalefcence, from the ftage of ftupor: with
thefe limitations, I think it a medicine poffeffed of
virtues beyond all eulogy.

Wine, fo univerfally prefcribed in low fevers,
as the moft grateful of all ftimulants, has frequently been condemned without reafon, or fallen
into difrepute, from a carelefs or injudicious exhibition. There are even phyficians at this day, fo
wedded to ancient theories of fever, as to limit the
quantity of wine in Typhus to three or four
ounces in the twenty-four hours; while others,
with unpardonable levity, are drenching their
patients to the amount of five or fix pounds in the
fame fpace of time. Thefe gentlemen, it may be
fuppofed, muft have occafionally been wrong:
medio tutiffimus ibis : and I fufpect, that neither of
them had fufficiently taken to their guide that
experience, which is to be gained only by very
frequent vifits to the bed-fide; becaufe the ftate of
the pulfe, and feelings of the fick man, are the
fure directors whether the allowance fhould be
fmall or large *.

It

* Erafiftratus primo tribus vini guttis, aut quinis afper-
gendam potionem effe dixit; deinde paulatim merum adji-
ciendum. Is, fi ab initio vinum dedit; et metus cruditatis
fecutus eft, non fine caufa fecit: fi vehementem infirmitatem
adjuvari poffe tribus guttis putavit, erravit.

Cel. L. iv. cap. 11.

It may be juftly afked, at what period of Typhus ought wine to be adminiftered, and by what rules are we to regulate the quantity?

To the firft part of the queftion I would anfwer: that wine may be ufed in moft cafes of this fever, from the earlieft acceffion, and in every period of its continuance. There are however exceptions, that ought to be cautioufly attended to. In fome conftitutions, and at particular feafons of the year, fome fymptoms of increafed excitement evidently continue for a few days. A greater degree of thirft, harder pulfe, flufhed countenance, higher coloured urine, are the chief appearances; at the fame time the abfence of thofe figns which more efpecially indicate debility, fuch as dejeftion of fpirits, and what are called nervous affeftions, &c. fufficiently fhow the difference. Under thefe circumftances, the good effefts of wine are doubtful; and what I confider a ftrong contra-indication, the patient takes it with reluftance; and it is followed by an increafe of the thirft, and flufh of the countenance. When thefe fymptoms difappear, the exhibition of wine becomes eminently ufeful: I fhall now mention how that may be determined.

The propriety of a gentle ftimulus to the body labouring under a difeafe of torpor and debility, I apprehend will be very generally admitted. But one that is grateful to the tafte, that communicates vigour,

vigour, and excites pleafurable fenfations; if fuch can be found in the materia medica, it will be the duty of the phyfician to prefer that. Such a ftimulus, and with thefe qualities, I confider wine. There are few medical gentlemen, I dare fay, who have attended patients in Typhus, that have not remarked the avidity for wine : the eagernefs with which it is received, and the inftant ftrength that it feems to infpire, are equally aftonifhing. I have frequently feen the cup broke between the hands of the fick man, in the earneft grafp to get it to his lips, or gnafhed between his teeth in the hurry of fwallowing it. I remember a feaman, belonging to the Valiant, very ill of the fever, after having finifhed his own allowance, feizing the bottle of his fhipmate, who lay next to him, drank the whole at once, and fell afleep immediately. Having accufed him next day of irregularity, and robbing his friend, he folemnly declared that he remembered nothing of the matter : he had no return of deli- rium from this time.   A black man, belonging to the Robuft, who lay fome days in a very doubtful ftate, but always took the wine with defire when it was put to his mouth, was afked, when the con- fufion of thought went off, if he liked his wine? replied with great emotion, " O, maffa, gib im a gallon." On entering one of the wards at Haflar, one of the Raifonable's people was fitting up in bed, with five guineas in his hand, and begged me

U   with

with great earneftnefs, to allow the nurfe to lay
the whole of it out for wine. This man had a
wavering about him at the time, but which was
always mitigated after his dofe. The natural
craving for this cordial, induced me to increafe
his allowance to four pints, in the twenty-four
hours; it was almoft the only cafe where my
prefcription exceeded three pounds, during my attendance at Haflar. He had been a week ill in
the fhip; there was a wildnefs in his looks not to
be defcribed, but it always denoted danger; he
died three days afterwards, and took his medicine
till within an hour of his diffolution. Hiftories
of this kind, fhewing the ftrong natural defire for
wine among typhus patients, might be extended
to a very great length; but this propenfity will
always be worth the attention of phyficians.

I have no doubt, but under particular circumftances, a greater quantity of wine may be proper,
and probably fomething may be determined by the
patient's general manner of living, which ought to
be taken into the account. It can be always with
fafety increafed, while the pulfe decreafes in frequency, or becomes more equal and full. Even
the countenance of the patient is fome criterion to
judge by; it partakes of that appearance which
we obferve in the focial circle at table; when the
exhilirating powers of wine flufh on the cheek,
and fparkle in the eye. When the lucid interval
of

of reafon continues longer, and the recurrence of
delirium prevented, by giving a glafs or two of
wine, we may be affured it is doing good.   When
the ftimulus has not been too great, but moderate
and fufficient, the fleep will be found, the breathing foft, and the pulfe more full.   Care, however,
muft be taken, that the fleep is not protracted fo
long as to endanger the finking of the pulfe, by
the ftimulus being too long withheld; it will
therefore be neceffary to wake the patient, in order
to repeat the medicine.   Even when he is awake,
and lying quiet, the wine muft occafionally be put
to his lips; the tone of the voice will tell the propriety of the draught, or whether it will be taken
with pleafure.   It ought to be given from a veffel
with a fpout, fuch as a tea-pot; becaufe, for the
greater part of the difeafe, and when weak and
languid, the patient lies on his back, and is unable
to bear an erect pofture.   So many unaccountable
recoveries have come under my own obfervation,
that I am never led to defpair, while the wine is
taken down; for I have feen it received with
emotion, even during the moft hopelefs circumftances otherwife.

The wine ought not, however, to be always
withheld, when the patient feems to have an
averfion to it.   The acidity of fome wines will
often produce this effect; and when the excefs of
acid in the ftomach has been corrected and

U 2   neutralized,

neutralized, the averfion will be overcome, when a
wine lefs fharp may be reforted to: a change of
wines ought, however, to be tried, before their ufe
is entirely laid afide. This diflike often affords
an opportunity for employing malt liquors, inftead
of the others. I have frequently feen porter or
beer preferred to wine, and always found advantage
in fatisfying the cravings of nature, and allowing
the patient what he liked beft. Bottled ale, that
undergoes a brifk fermentation on drawing the
cork, is one of the moft grateful drinks that we
know. After recovering from a low fever, which
I had caught from fome of my fchool-fellows,
when not quite twelve years of age, I was informed, that I had been fupported entirely by brifk
bottled fmall beer, refufing every thing befides
that was offered. It was in vain that the attending apothecary urged me to take his white
powders, which I fuppofe were the famous fever-
powders of Dr. James: by thefe means the effer-
vefcing fmall beer of an old lady in the village, got
the credit of my recovery. Sydenham has fomething to this purpofe. Cyder is alfo, on many
occafions, highly relifhed by fever-patients; its
natural fharpnefs and brifknefs, when poured
from a bottle, render it grateful. It is alfo lefs apt
to clog the ftomach than malt liquors, and as poffeffing a vinous fpirit, has moft of the advantages
of wine.

The

The practice of giving malt liquors, was introduced at Haslar while I officiated there. There
are many to bear testimony of its effects. It was
began at first, from a few of the soldiers belonging
to the 19th regiment, under Lord Moira, calling
for it with singular intreaties. About this time,
these men had shown symptoms of recovery; and
I consider this longing for malt liquor, as a pledge
of returning appetite. They had, probably, in
health, been accustomed to drink porter or ale;
it might, therefore, be trusted as a strong indication of those propensities, which are associated
from habit, with a state of health. At the same
time, there was not that desire for wine, which
they had expressed a day or two before; and on
this account, likewise, I was the more disposed
to indulge them. Two or three pints of strong
beer, were generally given in the day: the feelings of the sick, were the best proofs of its good
effects. They all agreed, that it did them more
good than any thing; because it was given at a
time, when the strongest desires of nature called
for it. This beer was brought from the tap-
house, at private expence; but Mr. Taylor, the
apothecary, furnished us, afterwards, with bottled porter, which is one of the best ingredients
in the diet of a convalescent, and never failed to
strengthen them quickly for duty.

U 3

In

In the Charon's hofpital, at different times, we have had two or three patients, to whofe tafte, both wine and malt liquor were difagreeable : yet rum punch was relifhed, and was taken in due quantity, with all the beft effects to be expected from wine : we ought, therefore, to vary the liquor to the tafte and defires of the fick, and never to defpair, while he will receive it in any form.

Nil defperandum, Baccho duce, et aufpice Baccho.

The late Mr. Kerr, furgeon of his Majefty's yard at Portfmouth, informed me, that when a very young man, and furgeon of the fhip, with Captain Bofcawen, afterwards the Admiral, he received the infection of this fever from the people on board. Having been moved into a cabin in the wardroom, for the benefit of pure air, extremely reduced and very low fpirited, he rofe from his bed, and in the abfence of his mefsmates, feized a bottle of wine from the table which they had juft left, the whole of which he drank. He flept fo long afterwards, that he was defpaired off ; but awaked in due time, refrefhed and invigorated in a manner, not to be defcribed. He had, during his illnefs, often afked for wine, which was as often refufed, under the idea that it would hurt him, from its heating quality. It was

much

much about this time that Sir John Pringle
gave it with fuccefs in the army.   Doctors,
Gregory and Cullen, at Glafgow and Edinburgh, entered fully into the practice: we alfo
find from Dr. Huxham, that it conftituted a
great part of his treatment in nervous fevers.—
Mr. Kerr added, that he perfifted in the ufe of
wine; and to it attributed his recovery from the
laft degree of emaciation and weaknefs of body.

Opium, next to wine, has been generally prefcribed in Typhus.   We no longer contend for a
fedative power in this medicine, in any other way
than as a ftimulant, exhaufting the fenforial power.
The fame fymptoms which indicate wine, alfo
evince the propriety of opium in this fever.   But
a phyfician who has been much converfant with
the difeafe, muft have remarked, that, under
certain circumftances, a perference may be fometimes given to the one over the other.   It ought
to be remembered that wine, as an article of diet
in health, and when ufed with moderation, is the
moft beneficial and agreeable cordial that we are
acquainted with.   Opium, on the contrary, with
us, is always an article of medicine, and never
reforted to but in a difeafed ftate of body.   While,
therefore, the human body is in fome meafure
accuftomed to the one, it has feldom experienced
the other.   There is, moreover, a nutritious quality
in wine, which opium does not poffefs.   Neither

U 4   of

of the two ought to be prefcribed, without the
phyfician attending by frequent vifits, to watch
the effects : but when opium has been given, this
is more particularly required. We have fpoken
of the defire which the patient expreffes for wine
and other liquors ; but opium has not been
equally acceptable, either to the palate or ftomach. Opium has peculiar effects on different
people : it is not eafy fo to regulate the dofe, as
to excite a due ftimulant power, without inducing delirium : I alfo think, that a greater degree
of debility follows ; both in the ftate of the pulfe
and in the ftomach, although the degree of previous excitement was apparently the fame. In
my practice, I have generally given opium at bed-
time, to affift the natural propenfity to fleep at
the ufual hour. It is from this effect, that I confider it a moft valuable medicine : but care fhould
be taken that the dofe may not increafe the debility we wifh it to cure ; or induce ficknefs and
vomiting, which it is apt to do next morning.
Many of the precautions we have given, when
fpeaking of wine, will apply to opium.

In urgent cafes of Singultus, Diarrhœa, or
whenever there is acute pain of any part, I have
recourfe to opiates.

Except in the Dyfuria of Typhus, I have not
witneffed much good effect from camphor. It
appears to me, from many trials, to be a ftimulant

2   not

not much fuited to this kind of fever. It alfo affects fome perfons very differently from others : I have feen it induce a degree of ftupor, truly alarming, in a young woman, by a moderate dofe.

I have already mentioned ammonia, favoura- bly in fome fymptoms of the early ftage ; and I have thought it ufeful in convalefcents when joined to the cinchona.

Æther, in fome fituations, I have feen ufeful. When there are any troublefome nervous fymp- toms, as commonly exprefled, as tremours, pal- pitation, &c. joined with opium, it acts power- fully in giving relief. To check vomiting, it is alfo of benefit. The fpirit of vitriolic æther differs little from æther itfelf, and may be con- fidered a weaker ftimulant. For thefe kind of fymptoms I have alfo tried caftor, mufk, oleum animale, and afafœtida, and often with advantage. Pure brandy will fometimes check vomiting, when every thing elfe fails.

Some remarks have already been made on the practice of bliftering. In local pain, I muft ad- mit their utility, but cannot go fo far as others have done in its recommendation : but pain, in moft parts, is more quickly relieved by opium. When applied to the head in delirium, I have often been difappointed in blifters; but not fo much as to urge me to condemn their application altogether.

In

In this fever, though evidently attended with great debility, I have frequently feen cafes, where I had ftrong fufpicions of an inflammation of the brain or its membranes. It confifts, in a ftate of ftupor, from which the patient frequently ftarts up, with an appearance of ferocity in the countenance, a flufhing of the cheek, the eye rather turgid and inflamed. Although the pulfe leads fo little into the nature of the fymptoms, yet thefe appearances contra-indicate the ufe of wine and other ftimulants, which feem to render the fick man more reftlefs. What I have obferved to give much relief, is fhaving the head, applying cloths dipped in vinegar or ammonia acetata over the fcalp. It is probable, that the cold produced by the evaporation of the fluid, acts as a fedative, and diminifhes the action of the blood veffels.

It remains for me to fay a few words concerning Relapfes. They may be defined, a renewal of the difeafed actions, from particular caufes, before the body had been in condition to refift them. Cold and fatigue are the moft general caufes of relapfe. They muft often happen to the poor, among whom this fever fo generally prevails; more efpecially as the beft means for recruiting health, and gaining ftrength quickly, are beyond their reach. The fame obfervation applies to

the

the failor and foldier, with the addition of hard duty and expofure to the inclemency of weather. The continuance of bark, for a certain time, with nourifhing food, and in full quantity, to which ought to be added wine for a limited fpace, and then the ftronger malt liquors, as ale and porter, are the beft prefervatives of health: exercife and warm cloathing are a neceffary part of this plan.

The medicine required after Relapfes, muft be regulated by the general method of treatment in Typhus. The antimonial preparations will form part of the cure, fhould the difeafe be ufhered in by heats and chills: otherwife we muft have recourfe to what has been found ufeful in the fecond ftage.

Such has been our general practice in Typhus. If it has exhibited little that is new, it will at leaft teach thofe who have had lefs experience, fome cautions in the ufe of former medicines. They will, therefore, have the advantage of beginning where I leave off. It has been my ftudy to fee my patients often, in order to watch the effect of my prefcriptions, which I think to be more neceffary in this difeafe, than has been ufually thought: Let me not, now, be condemned by thofe, who will not give themfelves equal trouble to be informed. My laft appeal, muft be to facts collected at the fick bed, where I have patiently

and

and faithfully done my duty, undifmayed with
the dread of catching difeafe from my patient.—
How I have always efcaped infection, I cannot
tell : " But God tempers the wind to the fhorn
" lamb."

Mr. Kenning's Diary of a Fever that appeared
in the Invincible, in July 1795.

Early in July, fome gentlemen in the cockpit were taken ill with Fever : the fymptoms
were thofe of a mild remittent, and gave way to
the ufual treatment ; viz, emetics, antimonials,
and the bark. There were three or four that
complained, but were nearly well about the 9th,
when one of the failors was feized with chills,
head-ach, and naufea, with quick pulfe : he took
an emetic, and hauft. anodyn. h. f. '*.

July 10th. Another taken ill; the fymptoms appear more violent : he took antim. tartar, gr. ifs.
natron vitriol, ʒ fs. which vomited him and procured two ftools. The other patient is ftill
feverifh. Pulv. antim. gr. vj. ft. in bol. 4ta.
q. h. f.

11th. The patient who complained on the 10th,
is to-day worfe ; his ftomach is very irritable

---

* Mr. Kenning does not, with his ufual accuracy, mention by what means this contagion was brought into the fhip.

with frequent inclination to vomit.   Mift. falin. hora fomni. Tinct. opii. The other is nearly the fame. Repetatur bol. antim.—One added to the lift.

12th. Three complained this morning.   I now perceive that a Contagious Fever has made its appearance in the fhip ; and every precaution is taken to feparate thofe who are ill from the fhip's company.   Rep. med.

13th. Two added to the lift, in whom the attack is violent ; the fkin hot, pulfe quick, headach, naufea, tremour of the tongue when put out, and a fenfe of cold down the fpine.   Let each take an emetic : bol. antim. nocte.—Thofe firft taken ill are delirious.—Opp. emp. cantharid. inter fcap.   Let them drink plentifully of gruel, with Nitre. T. Opii. h. f.

14th. Four complained this day ; the attack much the fame as in thofe laft mentioned : there is no perceptible remiffion, but confiderable exacerbation towards evening.   They continue to take pulv. antim. and nitr. occafionally laxatives, and opiates at bed-time.   One in delirium is worfe to-day ; has a great averfion to any liquid touching his lips *; is fo uneafy as to make it neceffary to confine him in his hammock.

---

* Hydrophobia, a fymptom frequently met with in this fever.   T. T.

15th.

15th. One added to the lift. I find that the antimonial medicines do not produce any moisture on the skin, or abatement of the symptoms. They now begin to take cinchona in vin. rub. opium at bed-time, and wine and water for common drink.

16th. In general worse : petechiæ appear on the neck, back, and breast, of three ; two of whom were those taken ill on the 13th, the other on the 12th : two, who appeared to be well last night, were seized with frantic delirium about one o'clock : they were walking under the half-deck, the place appropriated for convalescents. I rather suspect they had grog from their messmates.— Cont. med.—Three added to the lift.

17th. Petechiæ appear on others : three still delirious. Repetatur cinchon. cum vino. porter or cyder for common drink, being plentifully supplied by the Captain. Those in delirium to have their heads shaved, and fresh blisters : the urgent symptoms in some are, difficulty of breathing and deglutition ; in one, subsultus tendinum : hæmorrhage from the nose, in one who was taken ill on the 11th.

18th. Two added to the lift : comatose symptoms prevalent with six : the one mentioned with subsultus tendinum, worse ; dilatation of the pupils ; in the evening low muttering ; delirium in four. Rep. med. difficulty of breathing relieved by the blisters ; the bowels in all are regular : one vomited the cinchona, gave him an infusion of

quassia,

quaffia, gentian and ginger in port wine : petechiæ appear on the neck and breaſt of one, that
was taken ill on the 15th; ſlight hæmorrhage
from his noſe, diſtenſion of the alæ naſi, with
exceſſive tremour of the tongue. T. opii h. ſ.

19th. Exceſſive debility prevails ; dilatation of
the pupils the ſame, with ſubſultus tendinum et
floccorum collectio.—State of the liſt this day,
nine very ill, and twelve in a favourable way.—
The caſe mentioned as worſt, is apparently verg-
ing to diſſolution, paſſes his urine inſenſibly.
Miſt. camphor and T. opii, to thoſe who are
worſt ; drink as before.—One taken ill.

20th. Nearly as yeſterday : one added. There
is a conſiderable degree of fatuity attending patients in this fever, though otherwiſe, not very ill.
John Brady, mentioned as being ſo ill yeſterday,
has been got out of his hammock, and well waſhed all over with cold ſalt water, rubbed dry, and
put into a clean bed ; took vin. aromat. fervid.—
William Tirry, another that has been very ill
for ſome days paſt, is conſtantly covered with a
greaſy ſweat ; his ſkin clammy, and a conſtant
flow of moiſture from his eyes ; paſſes urine and
fæces involuntarily. Miſt. camphoræ. At night
all have wine, porter, or cyder.

21ſt. John Brady appears better ſince the ablution yeſterday : Tirry is worſe, mouth black and
dry, his tongue conſiderably enlarged, a circumſtance common with many others : one that ap-

peared

peared perfectly well for three days, was airing
his bedding, and was attacked with a violent fpaf-
modic affection of his breaft, which continued
moft part of the afternoon ; he was relieved by
æther. vitrol. and opium. One of the convalefcents
was fuddenly feized in the evening with fevere
pains of his thighs, legs, and arms, fo fevere as to
make him cry out. Opii gr. ij. Comatofe fymptoms ftill continue. Tirry is worfe this evening,
hiccup. Mift. Camp. et T. opii. Deafnefs in one,
the firft that complained. Rep. cinchona cum
vino, to all who can take it :—one added.

22d. Thofe juft taken ill have little appearance
of fever; pulfe nearly natural, and fkin moift; they
complain of partial head-ach, and general uneafinefs : where there is no naufea or coftivenefs, give
the bark in wine immediately, otherwife, an emetic, and if a ftool is not procured by that means,
natr. vitriol ʒi.—John Brady much worfe towards
evening, low muttering, delirium, and groping
about, his eyes conftantly turned up, and the pupils more dilated than before : Tirry's hiccup continues, with convulfive twitchings of the mufcles of
his face, mouth quite black and parched, and cannot
articulate. Vin. aromat, &c.—The firft taken ill is
conftantly covered with a recking or fmoking fweat,
his tongue moift and enlarged, has the appearance,
as if macerated in water. Mift. camphoræ et T. opii.
He that had the fpafm laft night, is well : the other
                                                    with

'with the pains, has a return this evening, and relieved by two gr⁵ of opium.—Six added to the lift.

23rd. Brady and Tirry infenfible; p. fcarcely perceptible : Brady died at eleven o'clock, the fourteenth day of his illnefs; Tirry at one o'clock, the eleventh of his illnefs: three others are very bad, one of whom paffes his urine, and another fæces, involuntarily, who has fub. tendinum with frequent ftartings, and apparently great opprefſion about the præcordia; app. emp. canth. fterno. T. opii and mift. camph. vin. aromat.— Peter Reid, mentioned as paffing his urine involuntarily, is delirious, p. quick and weak, will feldom take the bark; he is got out of bed and waſhed with cold falt water, and his bed fhifted, rubbed dry, and took vin. aromat. fervid; about an hour after defired to get up, made water, and had a copious ftool; at bed time a difficulty of breathing; emp. canth. fterno. mift. camph. et opii.—Five added to the lift. :

24th. Favourable appearances, except in one, who is infenfible; contraction of the mufcles of his face, which draws his mouth to the right fide; he died about noon, the ninth day of his illnefs; petechiæ appeared on him the third day, and a ſlight hæmorrhage from his nofe on the fourth. Peter Reid ftill delirious, though his fkin has a more natural. feel, and tongue moift; he will feldom take the bark, but drinks freely of wine,

X　　　　porter,

porter, or cyder.—Six added to the lift; the attack ftill more flight; an emetic, and afterwards cinchona in vino fecundo. hauft. opii h. f.

25th. Four added : John Crawford, the firft taken ill, continues to perfpire very much, and his deafnefs increafed; is very fond of cyder, of which he drinks freely; is conftantly fmacking his lips, as if tafting fomething. Peter Reid, as yefterday : hæmorrhage from the nofe of one of the others.—Rep. med.

26th. Favourable appearances : two added to the lift.

27th & 28th. Peter Reid is free from delirium, great debility, his appetite tolerable, will feldom take the bark; op. h. f.—John Crawford nearly the fame, thirft infatiable, drinks fmall-beer, porter, cyder, and occafionally cold water; deafnefs continues.—One added.

29th. As yefterday : two more added to the lift. The attack is now fo flight, as fcarcely to confine the patient, but ftill has the leading features of the fame fever.

30th. Recovering.—Two added.

31ft. Four added to the lift, in two of whom the attack feems to be more fevere than for fome days paft.—Peter Reid has a troublefome cough, takes an infufion of quaffia, gentian, and ginger, with port wine, and opiate at bed time.—John Crawford recovers of his deafnefs, continues to be thirfty.

§   Auguft

Auguſt 1. John Crawford recovers of deafneſs, and paralyſis of right leg and thigh: the others as yeſterday.—One added.

2nd. One of the women, wife to a man that had been ill, was ſeized this morning with fever; ſhe was fatigued with waſhing the preceding evening: ſhe had an emetic and opiate. Others better.—Two complained.

3rd. The woman ſtill feveriſh, great irritation of ſtomach. Magneſ. alb. in ſp. menth. and T. opii h. ſ.—Crawford's paralytic affection better.—Two added to the liſt.

4th. The woman better. Let her have cinchona e vino.—All the others recovering. This is the firſt day that we have had no freſh attacks.

5th. Some of thoſe who were worſt, troubled with cough, and a thin expectoration. Vin. tonic c. opii h. ſ.—One added.

6th. As yeſterday.

7th. Recovering.

8th. Coughs ſtill troubleſome.

9th. Four added to the liſt: one with ſwelling of the parotid glands, the others partial head-achs, &c.—Peter Reid, mentioned on the 23d ult. as very ill, has been recovering ever ſince; he is ſeized with a ſpaſm in the lower part of the abdomen; opii grˢ ij. The woman is recovered.

10th. Crawford is recovering from the paralyſis of his leg, deafneſs nearly gone, has a ſwelling of

his

his legs. Rep. med.—Peter Reid has a return of the spasm; rep. opium.

11th. As yesterday: the return of Reid's spasm is prevented by opium, previous to the time of attack.

12th. Four complained this morning: vin. tonic et op. h. s.

13th. Three added; one with swelling of the parotid: Crawford recovers but slowly.—The Scurvy begins to make its appearance.

14th. One complains of swellings of the parotid: others recovering.

15th. Crawford continues rather weak, though with a tolerable appetite.—Received refreshments from the hospital ship.

16th, 17th, and 18th. Every appearance of fever disappeared. A few in a convalescent state.— Scurvy gains ground very fast.

19th. Sent Crawford to the hospital ship, in a convalescent state.

———

It has not been thought necessary, in the above Extract, to mention the time of discharge, of the different patients from the sick list: after the 25th of July, they were nearly in proportion to those who complained.

Dr. Blane mentions, in his observations on the infectious ship-fever, that " Petechiæ only appear " in the latter stage of the disease, and in cases " of considerable danger." In respect to petechiæ,

chiæ, it was otherwife in the fever here mentioned, which I confider, ftrictly, fhip-fever.   In
every cafe in which they appeared, it was at an
early period, generally on the fecond, third, or
fourth day, and never after the feventh.   There
is no doubt, but there may be confiderable danger,
when petechiæ appear; but feveral patients that
were covered with them, recovered : two of the
three that died had them.

The three patients mentioned with fwelling
of the parotid glands, had not the leaft fever.   I
was rather alarmed at that fymptom occurring
in the decline of the difeafe in the fhip; as Dr.
Lind mentions it as a proof " of a violent con-
" tagion, though unattended with fever ; not-
" withftanding that, fuch as were in this manner
" feized commonly died."   Thefe men were not
confined, and recovered in four or five days. The
fwelling difappeared gradually, in the fame manner that I have feen in cafes of cynanche poroti-
dœa, though without any affection of the teftes.

There was every precaution taken to feparate
thofe that were ill from the reft of the fhip's
company.   The half-deck was fitted for the convalefcents ; which place, with the fick berth, were
occafionally purified with wood fires, and kept
very clean.   The cloaths of the fick were hung
up in the berth, at the time of fumigation, and
carefully aired afterwards.

It may appear rather extraordinary in a fea

X 3   journal,

journal, that the fick, after the fhip had been feveral weeks at fea, fhould be fupplied with cyder and porter of the beft quality, and alfo frefh provifions. But fuch was the humanity and liberality of Captain Pakenham, although his ftock was large, he left off the ufe of porter and cyder at his own table, that they might be plentifully fupplied; of thefe articles they were very fond, and ufed them in great quantity. He had, previous to our failing, purchafed a quantity of onions, for the ufe of the fick; which, with barley and other vegetables, were boiled with portable foup, and a piece of frefh mutton, which made an excellent dinner, to thofe ill, fometimes to the number of forty. Thefe fuperior advantages fecured the cure, and prevented relapfes.—— As a further proof of the good effects of the above regimen, in the fubfequent month, when moft part of the fhip's company were ill with Scurvy, thofe who had been in the fever, were in no inftance tainted. In obfervations on fea practice, I think it an object of confequence, to mention the advantages or difadvantages of diet.

(Signed)

T. KENNING, Surgeon.

Invincible, Aug. 30, 1795.

MEDICINA NAUTICA.

A G U E,

IT is not eafy to trace thofe circumftances, which
occafionally convert an Intermittent or Remittent Fever, into a continued Type ; and *vice
verfa.* There muft, however, be fomething in
their difpofition very much alike ; for wherever
we find Typhus affecting a number of people at
a time, we alfo find cafes of the Remittent and
Intermittent form. They occur particularly towards the decline of the contagion ; and I have
looked upon them always as harbingers of its
fpeedy extinction.   In the fever which fpread on
board the Vengeance, in January 1793, and continued among the people, more or lefs, for four
months, were a number of regular intermittents.
There was alfo a large proportion of Agues,
among Lord Moira's foldiers, who came to Haflar hofpital in January 1794.   In moft of our
infected fhips after the 1ft of June, fimilar forms of

X 4   fever

fever appeared.   I could find nothing uncommon
in the conftitutions of the people to modify the
action of contagion to thefe fingularities, nor can
I offer a plaufible conjecture on the fact.   Does a
weaker degree of contagion, produce Remittents
and Intermittents ?   Cafes of this kind, have
given birth to the idea, that Agues are infectious.
Cleghorn and Clarke, both high authorities, affert
this : yet I have never feen one Ague produce
another, where the original one could be traced
to communication with the effluvia of marfhes;
either in this country, or between the tropics.
Solitary inftances of Ague, are very frequent in
King's fhips.   During the late war, while the
Berwick was employed in the North Sea fquadron, Agues occured twice among a confiderable
number of the fhip's company.   The action with
the Dutch Fleet, on the Dogger Bank, happened
on the 5th of Auguft 1781 ; the men that were
badly wounded, with a few fick, were fent on
fhore, to very indifferently-conducted fick quarters, at Sheernefs.   The fituation of this place,
furrounded. by marfhes, renders it fubject to intermittents : our people were, to a man, feized
with Agues, either on fhore or immediately after
coming on board.   Some of them were very obftinate, and one or two terminated in dropfy :
yet not a man in the fhip was affected, but thofe
who had been at fick-quarters.   The feamen who
were

were but flightly wounded, were fent to the hofpital fhip which lay in the harbour, all of whom,
returned cured of their wounds, and in perfect
health. This hofpital fhip lay but a fhort diftance from the fhore, but fufficient to prevent
any bad effects from the wind blowing over a
marfhy foil *. Circumftances very much like
the preceding, happened in the fame fhip, in the
Downs, a few months after. The Agues were
brought from Deal Hofpital; a place, at that
time, every way calculated to generate contagion,
and lengthen difeafe. How often did I regret, at
that time, that I poffeffed neither rank or authority in the fervice, to correct fo depraved a fyftem
of medical difcipline! I fincerely wifh we may
hear better accounts from arrangements that have
lately taken place.

The intermittent fevers which I have met with
on the coaft of Africa, were confined to the feamen
employed in the boats. I never had occafion to
ufe the lancet in any of thefe cafes; but an emetic,
or a purge, were generally given early, and always
with advantage. In fome, the bark was tried
during the whole of the apyrexia; and in others, it
was given in half-dram dofes, every half hour,

---

* Other fhips having fuffered by the people getting intermittents at Sheernefs, ought to have been a reafon for difcon-
tinuing fick-quarters at that port, and trufting the whole to
hofpital fhips commodioufly fitted for the fick.

for

for four hours before the expected paroxyfm. The
refult was uniformly in favour of the latter practice. When the bark was given in large dofes,
from the termination of one paroxyfm to the beginning of another, before the rigours commenced,
it frequently happened, that the ftomach was over-
loaded, and ficknefs and vomiting were inevitable.
By thefe means, the fucceeding fit recurred with
greater violence, and the difficulty of curing
it was increafed. On the other hand, when the
cinchona was exhibited a few hours before the
acceffion of the paroxyfm, fix drams appeared to
be more effectual than eighteen taken during the
whole intermiffion, in checking the recurrence of
the difeafe. After it was ftopped, it was neceffary
to continue the medicine at the expected period
of acceffion, and to diminifh it gradually. Only
one cafe occurred that did not yield to this mode
of treatment. It was a relapfe, from improper
conduct on the fide of the patient : he left us in
the Weft Indies, and I am doubtful that he ever
recovered.

Among the Earl of Moira's troops at Haflar,
fome few Agues were very difficult of cure. The
arfenic folution, in every cafe in which it was given,
failed to give relief: the cinchona after it, was
adminiftered with fuccefs. Thefe foldiers were
extremely debilitated, and had fuffered much from
want of diet fuited to their diftrefs, as well as the
unwholefome

unwholefome quarters of a crowded tranfport. There was a neceffity for allowing them wine in great quantity, and a full diet of animal food to thofe who could ufe it. Under this regimen, it was furprifing to fee with what rapidity the cure was completed in fome inftances. Relapfes only occurred, when the wine was withheld contrary to my directions : in the cafe of a grenadier, no lefs than three relapfes happened from this circum-ftance alone. He was a tall flender man, and was much reduced by the difeafe ; but he had a good appetite in the abfence of the fit, and could eat his pound of meat in the day. There are fome medical attendants of hofpitals, that think a failor or foldier are not much entitled to fuch delicacies of diet, as a pint of wine. Thefe are gentlemen that ftep eafily into their appointments, who have never witneffed the hardfhips of a fea-life, or mi-litary fervice ; and thus they forget that fympathy that forms one of the fineft *traits* in the medical - character.

It has been already mentioned, that Agues became very frequent in the Vengeance towards the decline of the contagion. Many of thefe cafes happened at fea ; and from having them fo much under my own eye, I refolved to try the full effects of opium in preventing the fit. The greater part of my patients were Irifh landmen, lifelefs, timorous beings, and indolent to an extreme, when indif-pofed.

poſed.   The moment they felt the firſt approach of·the fit, they were ſure to run to the cockpit for relief.   The doſe of opium was generally ad-miniſtered at the door of my cabin, for there wa● ſeldom occaſion for them to go to bed in the ſick-berth.   This was done, either under my own di-rections, of thoſe of the firſt mate, Mr. Peter Blair, in whoſe abilities and attention I had much cauſe to confide.   If the firſt doſe did not bring on ſome warmth in the ſpace of ten or fifteen minutes, from twelve to twenty drops were given in the ſame manner.   The changes of the pulſe and feelings of the patient were often carefully watched. We never gave leſs than thirty drops the firſt time, and never needed to go beyond ſixty in the ſpace of an hour, for in no caſe did it fail to give relief in this time.   In a few minutes from the exhibition of tinct. opii, an exhiliration of ſpirits was perceived, which was quickly followed by a relaxation of the ſurface, the countenance looked cheerful, and and a fluſh was ſpread on the cheek. The pulſe from being weak, quick and ſometimes irregular; became leſs frequent, full and equal : an agreeable warmth was diffuſed over the whole frame, and every unpleaſant feeling vaniſhed, ſome-times in a quarter of an hour.   Sleep now and then followed a large doſe, but generally this did not happen.   As ſoon as any ſymptoms indicated another paroxyſm, whether on the following day,

or

or not till the tertian interval, the tinct. opii
was repeated in the fame manner as directed in
the former fit, and always with equal fuccefs, fo
that the patient feldom experienced much tremor
or fhaking. The fecond paroxyfm was commonly an hour or two later in the day than the
preceding one; and but few inftances were met
with, where any indifpofition indicated a third
attack at the expected period of acceffion. The
patients themfelves were not a little furprized at
the fudden change of their fenfations, by fo fmall
a quantity of medicine : they were certainly the
compleateft cures that ever came under my obfervation, and may juftly be faid to have been
effected, *certe, cito et jucunde.*

During the intermiffion, thefe men were fupported with nourifhing and ftimulant food, fuch
as could be procured on board. They had a pint
of found wine per diem, and fometimes more if it
was required. From my ftatement of the fever
which prevailed in the fhip, the Lords Commiffioners of Admiralty had bountifully ordered us
to be fupplied with fix pipes of wine, and it was
the fafety of many. Thofe who were in need of it,
had a piece of frefh mutton from the table of the
Captain or Officers; often a flice of pudding, and a
bafon of broth, and conftantly a fupper of flum-
mery, with a little wine and fugar. To the portable broth, which was always boiled for dinner,

was

was generally added a piece of mutton, sent by their Commander, now Vice-Admiral C. Thompson, who never had a sheep killed for his own use, without ordering a share for the sick. By these comforts of diet, some advantage might have been gained by medicine, that could not otherwise be expected: and it shews how feelingly our Officers attend to the afflictions of their people.

A case of intermittent fever, of the tertian form, was cured by wine alone, given immediately on the beginning of the rigours. The wine was administered in four-ounce doses, by Mr. Peter Blair, every ten minutes; there was a considerable degree of tremour and shaking, that did not quite leave him till after the fifth dose; so that twenty ounces of wine seemed to have the same effect as forty of the tinct. opii. There was no apparent intoxication; but sleep was speedily induced, and lasted three or four hours, when he awaked perfectly well.

A few intermittents appeared in the same ship's company, at Prince Rupert's Bay, Dominica; where we went to compleat our water after the unsuccessful attempt against Martinique. The ground at the foot of the hills which surround that bay, is low and swampy: the whole is covered with trees or brush-wood. It must occasionally be an unhealthy spot, and the source of those diseases which in certain seasons originate from marsh effluvia.

effluvia. Some officers were imprudent enough to allow tents to be erected for their people to sleep on shore; I believe no difcafes were the confequence; but that feemed to depend on the short time of getting their casks filled : it was not a practice to be imitated. Our method of cure was chiefly opium exhibited in the manner related above, and it was equally fuccefsful as in the former cafes. There were fome that required emetics, and feemed to approach more to the continued type. We had alfo fluxes at this bay; the people got new rum at the watering-place, and fome of them nearly expired from intoxication : this made many of them fall afleep on the ground, which was damp, and checked perfpiration.

A practice in Agues, nearly refembling the above, has been fuccefsful in my hands for fome years : there are certainly, however, intermittents to be met with, where bleeding and purging are indifpenfable; and the ufe of both opium and bark will be found more effectual after thefe evacuations. In ftrong robuft young people, who have lived in the country, this remark will particularly apply, and it ought to be attended to.

" Glenmore,

" Glenmore, Sheernefs, 10th Sept. 1796.

" DEAR SIR,

" A new mode of curing intermitting fevers,
having lately come under my obfervation, I take
the earlieft opportunity of requefting your opinion
of it : I mean by the application of tournequets to
the upper and lower extremities. A very ingenious young man, Mr. Kelly, late furgeon of the
Iris, has, I believe, publifhed fome account of it
in the Edinburgh Medical Commentaries, which
however I have not feen. Upon firft hearing of
it, I refolved to make the experiment the earlieft
opportunity, efteeming it a very fafe one, under
cautious management. Since I joined the Glenmore, I have had two patients in that difeafe. I
carefully watched the acceffion of the paroxyfm,
and immediately applied a tournequet to the op-
pofite leg and arm, and kept them on for about
fifteen or twenty minutes; taking care to unfcrew
them as occafion feemed to require. This procured a very effectual remiffion. In one of the
patients the fit was violent when the tournequets
were applied : and when they had remained on for
about ten minutes, he called out that he was quite
well ; and by continuing it a little longer, the fit
went

went entirely off. In the mean time, the ufual remedies were freely ufed, fuch as bark, wine, &c. Thus, by carefully watching the difeafe.a few days, its progrefs was completely put a ftop to; and a flight debility only remained. This is the only opportunity which I have had of making the trial; but I have little doubt of its efficacy in the difeafe.

" I can account for its effect only in this way; allowing the paroxyfm to be the confequence of fome derangement in the fyftem, by debility, the application of the tournequets, in the manner I have mentioned, may caufe a more abundant and fpeedy return of blood to the heart; and by that means remove, in fome degree, the proximate caufe of the difeafe, by giving a temporary ftrength to the animal powers.

I am, DEAR SIR,

Yours moft fincerely,

(Signed)

Thomas Grey.

To Dr. TROTTER,
    Phyfician of the Fleet.

Y

MEDICINA NAUTICA.

THE YELLOW FEVER.

THE ravages which this fatal Difeafe have made, during the prefent war, in our fleets and armies, are beyond all precedent : the infidious mode of attack, and the rapid ftrides by which it advances to an incurable ftage, point it out as one of the moft formidable opponents of medical fkill. It has offered the fevereft obftacles to military operations, which the hiftory of modern warfare can produce : the chief victims of its fury being the young, the robuft, and thofe free from other difeafes; by which means a campaign in the Weft Indies is now confidered as little better than *forlorn hope.*

While this mortality is going on among our troops in the Weft India Iflands, two authors, nearly at the fame time, have publifhed opinions on this Fever, and related a method of cure very much alike, though at a diftance and unknown to each other. It has been held out in thefe publications,

lications, that the danger of the difeafe was overcome, and a treatment eftablifhed, as fimple and
effectual as in a common remittent.   The firft of
thefe works has been greedily received by the medical world. A peftilential fever appeared at Philadelphia in the month of Auguft 1793, the nature and treatment of which, excited no uncommon difputes among the Phyficians; which,
at fuch a time, could not fail to increafe the
affliction, as it wrefted from the fick bed one of its
laft comforts, the hope of recovery by medical aid.

During this awful vifitation, Dr. Rush appears
like a faving angel, arrefting the arm of death;
and we are at a lofs, whether to admire moft, the
profundity of his profeffional fkill, or that vigour
and benevolence of mind, that bore him up againft
the dread of infection, and the calumny of his
cotemporaries : in both, he has given a valuable
leffon to Phyficians, and will hand his name to
pofterity, as one of the greateft models for imitation.   To have fled from the fcene, and quitted
the duties of the profeffion at fuch a crifis, would
have been bafe and unmanly ; but to fee his
opinions triumph over ignorance and obftinacy, is
a victory only worthy of a benefactor of mankind.

Dr. Rufh fuppofes this fever to have arifen from
damaged coffee, that was left to rot on the wharfs,
and from which noxious exhalations were fpread
that firft affected the neighbourhood, and afterwards more diftant parts of the city.   A certain

Y 2

ftate of the air is however admitted to render this *miafma* fufficiently active. We are apt to believe, that a peculiar conftitution of the weather, was adequate to explain the origin and propagation of this fatal epidemic, without including the influence of putrefying coffee, which at beft is problematical. Something ought, therefore, to have been premifed on the topography of the capital of Penfylvania, which we are furprifed to fee omitted, while fo many inferior circumftances are minutely detailed. Is not Philadelphia built on a low plain on the banks of the Delaware*?

The practice of the Sydenham of America, was bold and decifive, in this fever. It was one

of

* Since writing this accouut, a Captain of a man of war, juft returned from the Jamaica ftation, informs me, that much about the time when Dr. Rufh took notice of the putrefying coffee on the wharf at Philadelphia, feveral veffels laden with the fame produce came to Kingfton from St. Domingo. During the diftracted ftate of that colony, this article, with other productions, had been allowed to fpoil and ferment; the evolution of a great quantity of fixed air, or carbonic acid gas, was the confequence; and in thefe veffels, when opening the hatchways, fuch was its concentrated ftate, that the whole of the crew, in fome of them, were found dead on deck. A pilot boarded one in this condition, and had nearly perifhed himfelf. It does not however appear that a Yellow Fever was generated at Kingfton by this coffee: we muft rather fuppofe that thefe people died from breathing the carbonic acid gas, the very fame kind of air that frequently proves fatal in the wells of fhips. This fubftance might, no doubt, affift the pro-

grefs

of thofe opportunities, which nature now and then prefents, as a field for the difplay of fuperior talents, to fome favoured genius; where he burfts afunder the fetters of fyftem, to expand, at large, in a new fphere; which he peoples with new ideas, and new opinions, peculiar to their Creator. Who would not travel through this vale of tears, amidft blafts of contagion, to fhare the well-earned fame of Dr. Rufh!

The other publication on the Yellow Fever, is from the pen of Dr. Chifholm. The fever which he relates appeared at Grenada in 1793, while he was furgeon to the hofpital of the royal artillery at that ifland. The Doctor, however, confiders this fever as a new difeafe in thefe regions, and differing effentially from the fynochus icteroides of the country. Having given his arguments in favour of this diftinction, he next attempts to prove, that it was an imported contagion; and was brought from the coaft of Africa by a fhip called the Hankey, belonging to the fettlement of Bulam; " where people had been induced to " fettle, more from the delufive profpect of " wealth held out to them, and the *fanatic* " *enthufiafm* for the abolition of the flave trade of " the moment, than by the deprivation of

Y 3 " the

grefs of contagion at Philadelphia, becaufe it would render the atmofphere lefs fit for refpiration; a phyfician poffeffing a talent for inveftigation, like Dr. Rufh, could not therefore overlook its effects.

" the means of fubfiftence in their own coun-
" try *."

This fever, according to Dr. Chifholm's account, had, among its peculiarities, fome of the
moft

---

* Chifholm on the Grenada Fever, p. 83. The expreffions
here ufed, partake little of the liberal and enlightened fpirit of
other parts of the work. Why fhould a Phyfician, the pre-
rogatives of whofe profeffion are to alleviate pain, and prevent
the evils of human nature, accufe a nation, impelled by a
religious principle to put an end to the fale of fellow creatures,
of a fanatic enthufiafm? When a man appears an advocate
for the flave trade, does he not, by that, difavow religion, and
mock revelation? If we, like the French, fhould decree
death to be an eternal fleep, fuch a practice would be confiftent; and we might fcourge, pinch, and torture the flefh of
the African to the laft drop of blood, for the fake of gain.
But the dread of a *hereafter*, muft conftantly keep alive this
inquiry, till the decrees of heaven, and the laws of juftice, are
fatisfied.   When I vifited the iflands in 1793, I was glad to
find a general improvement in the condition of the flaves,
which had folely rifen from the difcuffion of the fubject at
home: nor did I find a captious fpirit among the planters,
who could reafon on their fide of the queftion with a becoming candour.   I was, with twenty Surgeons on the Navy
Lift, at the end of laft war, obliged, from neceffity, to feek
employment in the African trade, becaufe the Navy Board
refufed us fhips, and we had no half-pay.   Some of the number died, and the greater part of the remainder returned broken
hearted.   I communicated to Mr. Wilberforce what I knew
on the fubject; fome fpoke with indifference of my evidence,
becaufe I had been but one voyage.   Yet on this circumftance
I refted its validity: had I engaged fully in the trade, the love
of money might have entrapped my judgment, as I poffefs
frailties

§

moſt prominent charaƈteriſtics of the Endemic Cauſus of Dr. Moſely; viz. its affeƈting new-comers, and being almoſt confined to whites: from theſe circumſtances we are led to believe, that it was nearly allied to the Yellow Fever.

While the Vengeance lay at St. Kitt's, on the 31ſt of July 1793, with Admiral Gardner and the homeward bound convoy, I was requeſted to viſit a gentleman on board a merchant ſhip from Grenada, who was reported to have the plague fever, then raging in that iſland. I found him in a helpleſs condition, with many ſymptoms of the laſt ſtage; ſuch as, univerſal yellowneſs of the ſkin, black vomit, and convulſions: I preſcribed for him without expeƈtations of ſucceſs, and ſup-poſed that he would die the ſame evening; the Fleet ſailed next morning, and I heard no more of him. In the paſſage home, on the 22d of Auguſt, lat. 33:44°. long. 57:02°. therm. 81:0; a ſhip loſt her foremaſt in a ſquall of wind, and received other damage, when the Admiral made the ſignal for the Vengeance to take her in tow. The ſhip proved to be the Hankey from Grenada and

Y 4

Bularn.

---

frailties like all mankind. I ſhould not have touched on this buſineſs, was I not deeply affliƈted at ſeeing a man of great medical abilities, and general ſcience, ranged on the ſide of a traffic in human beings; there is not another phyſician in tha liſt but himſelf: it will go ſome way in depreciating the eulogy beſtowed on the profeſſion by Dr. Johnſon, if Dr. Chiſholm does not recant his expreſſion.

Bulam. Captain Thompſon ſent carpenters on
board, with the neceſſary ſtores to aſſiſt in repair-
ing her loſſes: they remained for three or four
days; but no ſickneſs followed, nor had there
been any perſon indiſpoſed ſince they left St. Jago.
On our arrival in England I was not a little ſur-
prized to find, that very particular orders had
been ſent from the Privy Council, or Secretary of
State's Office, to Liverpool, ordering the ſhip
Hankey under a ſtrict quarantine, left the Bulam
fever ſhould be communicated by the intercourſe.
This was done in conſequence of the information
from the governor of Grenada.

Theſe circumſtances made me curious to read
the account given by Dr. Chiſholm. Having
peruſed it, with admiration of the author's abilities, and the ſucceſsful, though ſingular mode of
treatment, I frequently made it the ſubject of
converſation with medical gentlemen lately returned from the Weſt Indies. Dr. Chiſholm
tells us, that the ſhips of war on the African
ſtation having ſent men to aſſiſt the Hankey, after
numbers had periſhed from the fever, received
the infection by means of this communication,
and that in the Charon thirty died, and fifteen in
the Scorpion. Captain Dodd, who at that time
had his broad pendant in the Charon, now com-
mands the Atlas of ninety-eight guns in the
Fleet: Mr. Smithers, the ſurgeon, is at preſent
in the Formidable, a ſecond rate, alſo in the Fleet:
from

from them I have copied the following narrative of their tranfactions with the Hankey.

When the fquadron under Commodore Dodd came to St. Jago, in 1793, the Hankey lay there, in great diftrefs for want of hands, having buried above one hundred perfons, men, women, and children, from the time fhe had been at Bulam. The fever was now overcome : Mr. Smithers faw two men that had lately recovered. He prefcribed to the mafter, who was ill of a venereal complaint, and for which he left him fome mercurials, with directions how to ufe them; at the fame time he left a quantity of bark. The Charon and Scorpion fent two men each, to affift in navigating her to the Weft Indies. The Hankey, at this port, was cleaned, wafhed with vinegar, and fumigated. No fever appeared in either of the men of war in confequence of this communication; they arrived at Grenada in perfect health, but did not go to the fame port of the ifland to which the Hankey went. The Charon at this harbour received fome feamen from the merchant fhips, then taking in cargoes for England; fhe had afterwards fourteen cafes of Yellow Fever, of which one died: but it is remarkable, that the Scorpion did not bury a fingle man during the whole voyage *.

It is probable, from thefe facts, that the Hankey did not import the infection that produced the

Grenada

(* See next page.)

Grenada fever; for, after the difeafe was worn out, fhe had a paffage to make to the Weft Indies of

many

" * Lift of Men who died on board his Majefty's Ship Charon, between the 22d of November 1792, and 4th of Auguft 1793, in a voyage to the Coaft of Africa, and Weft India Iilands.

Feb. 18. Ja⁵ Coglin - Flux. An old man very much broke down, and feverely troubled with a rheumatifm ; had done no duty during our paffage to the coaft ; was attacked with flux on the 18th of Feb. and died March 7th.

March 6. John Bean - Flux. A mate of a flave fhip that was caft away on the coaft, and in the laft ftage, was taken on board by the Commodore's defire, for a paffage to England ; died on the 22d of March, on our paffage to St. Thomas's.

March 10. Querra - - Flux. A black prifoner going to England to be tried for murder ; died on the 30th of March, at the Ifland of St. Thomas's.

June 22. Tho⁵ Dillon - Fever. A healthy ftrong young man, died on the 13th of July, being then on our paffage to England.

After leaving Grenada on the 4th of June, we had twelve or fourteen men attacked with the Yellow Fever; the infection was brought on board by fome impreffed men whilft at Grenada.

many hundred leagues. It is alfo doubtful how the effects left in the Hankey could produce the fever, for the bedding was thrown away, and what cloathing remained, had been aired, and, probably, had fcarcely been in contact with the body after being fick. Mr. Smithers was examined before the Governor of Grenada on the fubject, and gave his opinion, decidedly, that the Hankey did not communicate this fever to the colony: from our people remaining fome days on board, at fea, and efcaping with impunity, is a ftrong fupport to the evidence of Mr. Smithers. Dr. Chifholm has laboured his arguments to prove, that it was a new difeafe; though, perhaps, only the common endemic of the country, more aggravated by a greater number of raw Europeans being the fubjects of its influence, than he had been accuftomed to obferve before. This was the natural confequence of the war; for fhips are detained to wait for convoys, which would increafe the number of feamen on the fpot; at the fame time we find raw recruits had juft arrived from England. The pay of feamen is generally high on thefe occafions, and they would be better

enabled

nada. The difeafe fpun out to a great length, but all recovered except Thomas Dillon.

The above deaths, with one man killed on fhore at St. Jago, and another drowned on our paffage to England, were all that happened during our voyage.

(Signed)

J. Smithers, Surgeon."

enabled to indulge their debaucheries, which would finally expose them to the diseases of the climate. These circumstances taken together, sufficiently account for the variations of this Pestilence from the common Yellow Fever; and it comes near to what has been told me by some medical friends, who visited most of the islands during the mortality among the troops. As Dr. Chisholm has, therefore, been misinformed in some very essential parts of his inquiry, I cannot help thinking, that the account given by Commodore Dodd, and his surgeon, tends very much to shake the opinion of the disease being imported.

Dr. Chisholm also says, " that this fever was not confined to Grenada, but spread from thence to the other islands, and to the continent of America."   I think this assertion is given on too slight grounds; we have the authority of Dr. Rush, that the Yellow Fever of Philadelphia was generated on the spot.   Three young gentlemen died on board the Vengeance, a few days after leaving the islands, of a fever, with regular remissions, where a yellowness appeared about the fifth or sixth day; which I could trace to a swampy ground, where we got our water, a little to leeward of the town, at Monserrat.   It is of great consequence to the peace and security of society, that physicians should be particular in investigating the sources of contagious diseases; and

every

every candid mind ought to guard againſt any biaſs from opinions of theory. We have ſeen ſome very ill-founded alarms ſpread, from the idea that this fever might be imported into England by trading ſhips : a ſimilar dread was ſpread while the diſeaſe raged at Philadelphia. I do not think that the infection is in any danger of becoming active on this ſide the Atlantic; it ſeems directly oppoſite to the nature of Typhus, and they are ſubdued by a directly oppoſite ſtate of the atmo-ſphere *.

From the account given by theſe authors, we have a right to conclude, that the diſeaſes, which they deſcribe, were the ſame; or what variation appeared in the ſymptoms, might be juſtly con-ſidered as the effect of ſituation, or ſome peculia-rity of conſtitution in the perſon affected. In the chief part of the deſcription, theſe phyſicians differ little from Dr. Moſely, to whom we are indebted for the beſt hiſtory of this Fever, al-though he does not ſuppoſe it to be infectious. We have, therefore, the concurring teſtimony of the three lateſt writers on the ſubject, that the Yellow Fever is, in its firſt ſtage, a diſeaſe of a high inflammatory diſpoſition ; and, in its latter ſtage, as the conſequence of the preceding, at-

---

* The dread of importing this fever, is kept alive by mak-ing ſhips undergo quarantine. Nay, an *army phyſician* reports the ſtate of men of war, before they are liberated by the health officers !

tended with fymptoms of exhaufted energy, terminating in gangrene and putrefaction. They
alfo agree, that a peculiar ftate of body fayours
its attack; viz. the young, perfons of tenfe fibres
and rich blood, and the new-comers from northern countries. This being the cafe, the indications of cure have been fimilar; although
anfwered by fome means that are rather new to
the treatment of febrile difeafes.

The theory of this fever, if I may be allowed
the expreffion, affords a fine illuftration of the
new doctrine of Brown and Darwin, in what relates to the difpofition of new-comers from cold
countries, making them the chief victims *. We
are there taught, that animal exiftence is endowed
with a quality, called *Excitability* by Brown ; *the
Spirit of Animation,* or *Senforial Power*, by the author of Zoonomia ; on which the phænomena of
life depend. Every thing which fupports life,
exerts its influence on this principle : it is capable of different degrees of accumulation and exhauftion : it is accumulated by the fubftraction
of ftimuli, and is exhaufted by exceffive ftimuli,
or the long application of others acting more

* See the Tranflation of the ELEMENTA MEDICINÆ of Dr.
Brown, publifhed by Johnfon, London, with the elegant Bio-
GRAPHY of Dr. Beddoes.——ZOONOMIA, or the LAWS of
ORGANIC LIFE ; by E. Darwin, M. D. Johnfon, London.
2 vols. An ineftimable work ; and when phyficians fhall
take the trouble to perufe it, will probably fupercede all other
fyftems.

mode-

moderately : a due equilibrum between the exciting powers, fuch as heat, food, air, mental
exertion, &c. and the fenforial power conftitutes
health ; and every variation from either, is *difeafe*.
When the fenforial power is accumulated, it is
more fufceptible of ftimulants ; this is ftrongly
exemplified in froft-bit toes or fingers, on being
fuddenly expofed to the fire, or plunged into
warm water : the pain becomes intolerable from
inflammation, and mortification, with mutilation
often follows.   This effect of the inflammation, is
the indirect debility of Brown, and the exhaufted
fenforial power of Darwin.   On a larger fcale, for
the fake of comparifon, we may fuppofe the inhabitant of a cold country, like a froft-bit limb,
and the climate of the Weft Indies, like the warm
water.   The cold of our winter, from deficient
ftimulus, allows the fenforial power to accumulate ; and increafes the fufceptibility of the body,
for the action of all exciting agents : in other
words, it predifpofes us to inflammatory difeafes.
As we change our feafons gradually, we feel little
inconvenience, becaufe the exciting power of heat
in the fpring, is not much beyond the degree of
excitability which the winter had accumulated,
and thus the one is gently exhaufted by the other.
So a froft-bit limb is recovered by plunging it
into water of a low temperature, more proportioned to its degree of excitability.   On thefe
principles, inflammatory affections are found to

prevail moft in the fpring months, when the heat
fucceeds quickly to froft. But if we go in the
fpace of a few weeks, from the cold of a fevere
winter, when the Therm. was at 20. to the Ifland
of St. Domingo, where the heat is never below
80°. the tranfition is too fudden, and the difference of temperature is too great for the body to
accommodate itfelf to it, at once. It is like putting the frozen limb into hot water : there being
a redundancy of both excitement and excitability,
which paffes rapidly through a ftate of the moft
violent inflammation, that terminates in debility,
gangrene, and fpacelus. To wear out this accumulated excitability, by flow and gentle gradations, is the grand explanation of the word *fea-
foning :* it is the fecret, which conftitutes the only
difference between the inhabitants of England
and Jamaica. The Yellow Fever of the Weft
Indies, therefore, as it appears in the body of a
raw European, is a difeafe of the utmoft excitement, in a conftitution of accumulated excitability ; where a tenfe fibre, and denfe blood, permit it to be carried to the higheft pitch of inflammatory tendency, which, from the nature of
the animal œconomy, fpeedily exhaufts the powers
of life, even in the fpace of a day or two, inducing putrefaction and death.

Thefe principles being eftablifhed, concerning
the nature of this fever, we have feen them applied

plied with fuccefs in the cure, by Doctors Mofeley, Rufh, and Chifholm; and we fhall now endeavour to reduce them to fome practical advantage
in the prevention.

We know, from Dr. Rufh, that a ftate of the
atmofphere, firft checked his awful epidemic; and
we know, from Weft India phyficians and furgeons, that a quick paffage to the cold latitudes,
often brings about wonderful changes in the Yellow Fever. The cold in this cafe reduces the
exceffive excitement, and prevents it from gaining
that height, which tends to exhauft the excitability or fenforial power in fo fhort a fpace.
When, therefore, the fever fpreads in a fhip, it
has always been reckoned a falutary meafure to
put to fea; we are only furprized that it has not
been done more frequently, during the late mortality among feamen, at the iflands. A certain
height of temperature, feems alfo neceffary to
give activity to this contagion; if the fever is to
be extended by this manner.

It was a wife meafure in our Government, to
fend troops from Gibraltar, or our iflands in the
Mediterranean, to garrifon the Weft India Iflands,
rather than from England at once: by which,
the tranfition of climate was more gradual, and
the exceffive heat fuftained with impunity. To
fpeak more technically; the accumulated excitability was exhaufted by the exciting power of

Z.

heat,

heat, by gentler gradations, till it became proportioned to the excitement ; which would fix
the equilibrium between the two, and conftitute
Weft India health : the mufcular fibre would
become more foft and lax, and the blood lefs
florid and denfe ; and thus the body would be
accommodated to a tropical temperature, or
feafoned. But, as military operations have been
carried on in that country, in a more extended
fcale than formerly, which puts it out of our
power to feafon a whole army at Gibraltar, the
the next expedient ought to be, to fend them
from this country before the cold weather fets
in, that they may arrive at the iflands in the
coldeft months. I therefore, think, that all re-
inforcements ought to leave Europe, from the
20th of September to the 1ft of October, but
not later ; whether fhips or regiments.

There is another circumftance deferving confideration in this bufinefs. Our armaments have
lately fuffered much by detention in the Channel :
now, it would much facilitate all foreign expeditions, if the whole of the army were to embark at
Cork, or other ports in the Weft of Ireland, by
which means they would *clear* the land in a day
or two. I would, therefore, recommend, for confideration, the conftruction of barracks for ten
thoufand men near the Cove of Cork, to be in
conftant readinefs for embarkation when neceffary

fary. The fate of the fleet under Rear-Admiral
William Parker, in December 1794, and of that
under Sir Hugh C. Chriftian, in November 1795,
are recent proofs of the propriety of this propofal. With refpect to the tranfports themfelves,
there is much room left for improvement : it
would furely be a better method for conducting
this fervice, that Government fhould build veffels exprefsly for the purpofe, and not permit
human beings to be fqueezed into fmall veffels; which in blowing weather are drenched with
water, and fome of which have occafionally been
found fo bad, as to be in danger of finking at
their anchors. To prevent unpleafant alterca-
tions between officers of the navy and army, they
might be navigated as in the eftablifhment for
navy tranfports. To each, there ought to be a
furgeon or mate, with neceffaries, inftruments,
and medicines. Many recent misfortunes, fufficiently evince the neceffity of this addition.
But to our fubject :

The prevention of this fever, evidently depends
upon reducing the excitement. Now as we cannot reduce the heat of a Weft India climate, we
ought to attend to the other ftimulant powers,
which act along with the heat of the atmofphere,
and decreafe thofe that we can command ; meat
and drink, air, &c. are all of this defcription : we
can reduce the body very much by the evacuations of bleeding and purging. There was an old

cuftom of bleeding and purging failors and foldiers, when a fhip got into the warm latitudes. I
cannot help thinking, that it was a rational practice, however ill underftood, and I [long to hear
of its being revived. I have feen a very quick
tranfition of temperature produce grievous head-
achs ; for one of thefe I bled myfelf at the ancle,
and had immediate relief.   They are certainly
genuine precautions againft tropical fevers ; now
confirmed by medical authority, and ftill more
by the premature death of thoufands of our
countrymen.

Heat, at 84º of Farenheit, is an exceffive
ftimulus to our bodies ; but its danger is much
encreafed, when at the fame time a great quantity of animal food, with vinous or fpiritous liquors, forms the diet. Thefe articles, in conjunc-
tion, carry the excitement to the higheft point
imaginable : and hence, luxurious living has lately
made many Englifhman food for this mortal
diforder : I have frequently mentioned the names
of my acquaintance to their friends, as being certain victims, from thefe caufes.

The fagacious Dr. Rufh, in addreffing his
fellow citizens, fays, " the beft preventive of
" the diforder, are, a temperate diet, confifting
" chiefly of vegetables, 'great moderation in the
" exercifes of body and mind, warm cloathing,
" cleanlinefs, and a gently open ftate of the
" bowels." I am afraid few of thefe rules have

been ftrictly followed, by either officer or private, in our army. As my refidence has been chiefly at Portfmouth for the laft eight months, I have had many opportunities of feeing and knowing the habits of life of a great part of the young men who embarked lately on Weft India expeditions. It was melancholy to behold fome of their meffes, where they feldom parted company, after dinner, till the bill would amount to a guinea, and often more, a-head. What a heedlefs profufion of health! what an example to men under their command! Thefe modes of living, I am forry to learn, were practifed by many of them throughout the paffage; and continued to be a cuftom in the iflands. Would not Government, in fuch cafes, have acted a juftifiable part, to have prefcribed a diet of temperance, and enforced the compliance, under a penalty of difmiffion? Many of thefe gentlemen had juft returned from a campaign in the Low Countries, where the moderate indulgence of wine was a fafe preventive againft the cold, and the marfh effluvia of the foil; but to carry fuch opinions and practice into a fultry climate, was certain deftruction. When giving my opinion to my friends, on the beft means of fecuring health in the Weft Indies, I often found it difficult to convince them of the propriety of low living. The common argument was, how are we to fupport the heat, unlefs well

Z 3 fupported

supported by nourifhment? This was exactly inverting the queftion : the heat is to be diminifhed by food lefs ftimulating. It feems never to have entered their heads, that the heat of the climate did, in a warm region, what a nourifhing and ftimulant diet affected in a cold one, to fupport the excitement. Intoxication, it is well known, is doubly hurtful in a warm country ; becaufe, with the fuperabundant ftimulus of the vinous fpirit, is joined the exceffive power of heat : the debility which it leaves behind, is more difficult to be recovered, becaufe the preceding excitement had been raifed to a very high degree. A dinner in the Weft Indies, where two-thirds, or three-fourths of it, has been made on vegetables, with water, fmall beer, or a little wine diluted, gives the body a degree of ftrength and activity, that are not to be obtained from a repaft of animal food, ftrong malt liquor, or wine. The latter bring on a drowfinefs and langour, that are apt to terminate in fleep, and which generally happens ; they are, therefore, improper food, while the heat of the atmofphere is fo great; they exhauft life by an unnatural ftimulus, and induce what is called *indirect debility*.

It is very generally admitted, that the French bear the climate of the Weft Indies, better than Englifhmen : now, no reafon can be given for this, but their temperance in living, or being natives

-tives of a warmer foil, in the fouthern provinces of France. With fuch an example before us, we ought to have done otherwife: *fas eft, et ab hofte doceri* : the maxim was never more applicable than in the prefent inftance. I cannot help thinking, for thefe reafons, that much might be done by diet and regimen, to preferve our failors and foldiers in thefe iflands.

As a large proportion of animal food enters into the diet of both failor and foldier, it would be neceffary to diminifh that, in the firft place : whether falt or frefh, it is the moft ftimulant diet. As foon as any outward bound fhip or tranfport, in its paffage, gets to the fouthward of the ifland of Madeira, let the beef and pork be reduced one half. In lieu of the other half, let the people be ferved with cocoa, coffee, or even tea, fufficient for a pint of water, with fugar to fweeten it, for an evening meal. If this does not equal the value of the beef or pork, let them be paid in cafh, as is ufual on fhort allowance. When they arrive at the iflands, let the daily quantum of meat be reduced to one third of a pound. If the failors have not oatmeal gruel, with fugar or molaffes for breakfaft, coffee or cocoa ought to be invariably fubftituted. A delicious mefs for breakfaft may be made of the peafe prepared in the manner of coffee, and fweetened with molaffes ; the people like it beyond any thing ; and I con-

Z 4

fider

fider it a highly worthy practice, and a moft judicious improvement in diet.　At this time, I would
alfo increafe the quantity of bread, which would
be always confumed, while they have cocoa, coffee
or tea, to encourage its ufe.　When the tropical
vegetables and fruits can be procured, I would
recommend them to be ferved, in harbour, two or
three times a week : veffels ought even to be conftantly employed to carry them to the fleet, at fea;
and if they cannot be fupplied in fufficient quantity at one ifland, let them be brought from
another.　Plantains and yams would be a fair
fubftitute for bread, at leaft once or twice a week.
Seamen and foldiers, it may be faid, would not eafily
fubmit to thefe changes : I well know they would
not be popular, till the men were convinced that
it was folely done for their good ; and it would be
the duty of every officer to imprefs them with this
truth.　For my own part, I fee no unfurmountable
difficulties in the bufinefs.　Many of the Weft India
vegetables, when mixed as a fallad, are delicious;
and a few ounces of falt meat are fufficient to relifh
fo many pounds.　But it would even be a cheap
method, to have thefe articles in conftant cultivation, for the ufe of navy and army, at every
ifland.　Ship loads of potatoes, and our cheaper
pickles, might be occafionally fent from England.
The flour pudding with currants, pea-foup, rice
and

and fugar, might alfo be judicioufly blended with
this diet.

The negroes live entirely on vegetable food, yet
they are capable of working for fixteen hours, out
of the twenty-four; which is a proof, that deficiency of animal food does not induce weaknefs
of body.   Vegetable food affords a larger quantity
of water, which dilutes the blood, gives a copious
fupply of perfpiration, during the evaporation of
which, from the furface of the body, the extreme
heat is leffened.

The ufe of fpirits, in any form, fhould be totally
laid afide : they are flow poifon in cold countries,
but here they do the work of deftruction quickly.
Four ounces of wine, forenoon and afternoon,
diluted with water, is an ample allowance for any
man, who wifhes to preferve health and conftitution in thefe regions : nay, I am of opinion,
that the man who drinks nothing beyond pure
water, has the beft chance of efcaping fevers in
the Weft Indies ; unlefs fome particular frailty of
body urges the propriety of drinking wine as a
medicine.   Among my own acquaintance, who
have lately returned in perfect health, I can reckon
many, who lived almoft on vegetables, and drank
water.   Some of the number confider themfelves
as living monuments of my advice, and attribute
their efcape to the inftructions they received from
me at their departure.   It is the effect of high living,
whether

whether in eating or drinking, to wear out the conftitution; it gives a fhort vigour, which is followed by langour and inaptitude to exercife: if in the laft ftate the body is expofed to unwholefome fwampy fituations, or the dews of the night, a fever follows. In conftitutions accuftomed to live in this manner, it riots with unconquerable violence; till in the fhort fpace of a day or two, this goodly frame is converted into a mafs of corruption. Thus we fee them the firft to fink under fatigue*, while the fpare and flender fubject undergoes it with impunity, and escapes fevers in all their fhapes.

By way of indulgence, a fhort allowance of malt liquor might be ferved once a week; but fpirits in any form fhould never be thought of. Uncommon pains ought therefore to be taken to diffuade men from the ufe of the rum: intoxication muft therefore be punifhed with the moft exemplary feverity.

The

---

* " He had a fever when he was in Spain,
" And, when the fit was on him, I did mark
" How he did fhake: 'tis true, this god did fhake:
" His coward lips did from their colour fly;
" And that fame eye, whofe bend doth awe the world,
" Did lofe his luftre. I did hear him groan:
" Ay, and that tongue of his, that bade the Romans
" Mark him, and write his fpeeches in their books,
" Alas! it cry'd *Give me fome drink, Titinius*;
" As a fick girl."

SHAKESPEAR.

The exemption of the fair fex from this dreadful malady, can only be attributed to the habitual and temperate tranquillity of their lives. Abftemioufnefs and moderation in eating and drinking, along with a fedentary life, give that delicacy and foftnefs to the mufcular fibres, and difpofe them lefs to difcafes of high excitement. Having lefs intercourfe with the hurry and buftle of bufinefs, that amiable equanimity of temper is feldom ruffled by thofe paffions which agitate mankind, and fupply an excefs of ftimulus. Bountiful Nature, has thus kindly attoned for thofe ills which are peculiar to delicate frames; while their virtuous habits give them that fortitude and patience, fo often admired under affliction, to which we are ftrangers. It is, therefore, women of more robuft conftitutions only, who become fubject to the Yellow Fever.

With refpect to drefs, I think fome changes might be made to advantage. I would have the whole cloathing of a foft, flight cloth, between a flannel and what is now in ufe. Socks and pantaloons, with canyafs fhoes, for the legs and feet: an eafy loofe fhirt or waiftcoat, of the fame woollen cloth, with fleeves, and no linen next the fkin. The foldier, for fleeping in, fhould be provided with a large blanket of the fame cloth, which he might occafionally carry with him, if on a march at night. For the head, I would prefer a light hat,

hat,, painted water-proof, and formed to defend
the eyes from the glare of the fun. Soap fhould
be allowed to wafh this drefs twice a week, at
Government expence. The hair fhould be worn
fhort.

A judicious and humane commander, whether
on board or on fhore, will always be careful how
he expofes his men to a meridian fun, in a warm
country. The duty will therefore be fo conducted,
as never to endanger them from the heat of the day:
it would be well if black people were to be employed, when it is neceffary to work in the fun.
A few negroes attached to every fhip and regiment,
would fave the lives of many white men. It is
furprizing to think, that this has never been accomplifhed, when we look to the expence and difficulty of procuring failors and foldiers. Amidft
the fluctuation of commanders, a progreffive fyftem of improvements will fcarcely take place : different men have different modes of action ; and if
ample documents of bad meafures point out the
neceffity of alteration, the beft concerted plans may
be laid afide, by a new fet of men coming into
office. But fuch is the want of forefight in human
nature, that the evidence of misfortune is feldom
canvaffed till too late.

An open ftate of the bowels is an excellent affiftant to health in the Weft Indies; it is indeed
indifpenfable, and fecures the ftomach againft a
regurgitation

regurgitation of bile, fo prevalent there *. To the precaution mentioned above, moderate exercife, regular hours of reft and fleep, with due attention to cleanlinefs, in perfon and cloathing may be added.

It, does not appear to me that much trouble would be required in introducing thefe alterations, in either navy or army : additional expence there could be none. I am aware that many objections may be offered; but they can only come from people who are torpid, and never thought of improvement. Let us abftract ourfelves a moment from felf, and think of our fellow-creatures. Look to the plains of Leogane and Baffeterre, fattening with the bodies of Englifhmen! Something muft be done to reconcile our countrymen to fervice in thefe iflands; otherwife, we muft abandon the lucrative commerce of thefe fertile regions.

There having been fo much novelty in the exhibition of mercury, by Doctors Rufh and Chifholm, in this fever, that we wait with anxicty to

learn

---

* Capt. Ball informed me, that the effence of fpruce a little diluted, did wonders at St. Domingo, both in prevention and cure. It acted as *a purgative*, and confequently was judicioufly adminiftered.—I have juft feen an advertifement in the news-papers, from the Commiffioners of Sick and Wounded, to contract for a large fupply of it. Capt. Ball added, that it fat on the ftomach when every thing elfe was rejected.

learn the general fuccefs of it among other phyficians. What information has come to myfelf,
on the fubject, has been very unfatisfactory indeed:
fome perfons condemn it, while others pronounce
it eminently fuccefsful. There is one fact, too
certain, that whole armies have died rapidly away.
We have therefore a right to conclude, that this
practice has either been of no effect; or elfe, it
has not been duly adminiftered. For the honour
of human nature, and the credit of the medical
profeffion, we cannot a moment harbour the idea
that it has not been duly adminiftered, or that
juftice has not been done to the mercurial procefs,
while the lives of fo many valuable men were depending on it, and the fate of a nation with mil-
lions of property finking with them. I would recommend a ftrict official return to be called, from
every phyfician and furgeon now in the Weft
Indies ; and a detail of the method of treatment
in the Fever, down to the minuteft circumftance
of medicine, diet, and regimen. This return would
no doubt exhibit a gloomy catalogue of mortality;
but it might go a great way to prevent a repetition
of the fcene. If accurate vouchers are required
for every two-penny nail, or yard of cordage, how
much more are they required for the fafety and
cure of a brave failor or foldier?

Might not fome method for generating artificial cold, be of fervice in this fever? I fhould
like

like to know the effects of wrapping the body, for a length of time, in wet linen; some how after the fashion of cooling wine, by suspending it in canvass bags, frequently sprinkled with water: such a trial is perfectly consistent with the most approved opinions on the nature of the Yellow Fever; but it must be done during the inflammatory stage.

———

September 12, 1796.

SINCE making these remarks on the diet of our ships, I find they have not that originality which I believed belonged to them. I have just met with a work, entitled "*Colloquia Maritima*, or Sea Dialogues; by N. Boteler, Esq. formerly a commander in one of his Majesty's royal ships: London, printed by W. Fisher and Richard Mount, at the Postern on Tower-hill, 1688." In which a proposal very much resembling my own, is narrated by the captain, in a dialogue with his admiral *.

" Admiral.—Let us now return to our victuals,
" wherein there is one point more that I desire
" to be satisfied in; and that, whether it were not
" more beneficial and preservative for the health
" of our men, that the main of our victualling
" were

———

* This curious work was put into my hands by my worthy friend Vice-Admiral Gayton, who preserves it as an antique,

" were, in the kinds thereof, altered and nearly
" fitted to the manner of foreign parts, rather
" than as at the prefent with us; to confift fo much
" of falt and powdered meats, in beef, pork, and
" falt-fifh ?

" Captain.—Without doubt (my Lord) our
" much, and indeed exceffive feeding upon thefe
" falt meats at fea, cannot but procure much un-
" healthinefs and infection, and is queftionlefs
" one main caufe, that our Englifh are fo fubject
" to Calentures *, Scarbotes †, and the like con-
" tagious difeafes, above all other nations; fo
" that it were to be wifhed, that we did more
" conform ourfelves, if not to the Spanifh and
" Italian nations, who live moft upon rice-meal,
" oat-meal, bifcake, figs, olives, oil, and the like;
" yet at the leaft to our neighbours the Dutch,
" who content themfelves with a far lefs pro-
" portion of flefh and fifh than we do; and
" inftead thereof, do make it up with peafe,
" beans, wheat, flower, butter, cheefe, and thofe
" white meats (as they are called).

" Admiral.—It were well indeed if we could
" bring ourfelves to this provident and whole-
                                            " fome

---

* A phrenfy or inflammation of the brain: it ufed to be
frequent in fhips in the Mediterranean and Weft Indies. Inflammation of the brain, and the ferocious delirium, are common attendants of the Yellow Fever.

† The Scurvy. Latin, Scorbutus.

" some kinds of sea-fare; but the difficulty con-
" sisteth, in that the common seamen with us, are
" so besotted on their beef or pork, as they had
" rather adventure on all the Calentures and Scar-
" botes in the world, than to be weaned from
" their customary diet, or so much as to lose the
" least bit of it; so that it may be doubted, that
" it would set them upon a loathing, and running
" away, as much as any other thing whatsoever.

" Captain.—I confess, that it is no easie matter,
" by any new reason, to take off these from an
" old custom; and yet would they but patiently
" consider, of the well and lusty subsistance of
" the Italian, Spanish, and Dutch nations, who
" hereby live far more healthfully at sea than
" they do; or but of our colony people in St.
" Christopher's, the Barbados, Virginia and the
" Bermudas, who for the most part live, and
" thrive well, with their *husked-homeny*, and *lob-*
" *lolly* (as they tearm it) which they may make
" of the West Indian corn, called maize, it would
" perhaps work them to some willing conformity
" in this particular; or if not, it is fit that they
" should be used like *little children*, or *peevish*
" *patients*, and made to keep a good diet
" whether they will or no*.—But howsoever

* I suppose this frank, though intelligent officer meant,
that if they would not comply with what was for their benefit,
and the duty they owed to their King, that they should be
either flattered or whipped into it.

" sure

" fure I am, that this maiz is a moft excellent
" fea-food, and moft proper for long fea-voyages;
" for (as it may be eafily ordered) it will keep
" extraordinarily, and withal is very nourifhing
" and healthful.

" Admiral.—But I fee not of what ufe it can be
" with us, fince it groweth not in thefe parts,
" nor is here any where to be had.

" Captain.—I know well (my Lord) that thefe
" northern climates produce not thefe kinds of
" grain; for neither the heats of our fummers,
" nor the ftrength of our foil, will bear or mature
" it; I do not therefore propound it as a pro-
" vifion for our fhips outward-bound, but only
" to intimate, that whenfoever we fhall have oc-
" cafion, and leave to look once again towards
" the Weft Indies, that then this kind of food
" may be found moft ufeful for a fupply of
" victual to all fuch of our fhips as are bound
" that way, and that either whilft they are there,
" or when they are to return. To which end it
" will then be neceffary that all our fouthern
" colonies be inftructed to employ themfelves,
" (rather than as at the prefent upon *fmoaky*
" *tobacco*) in planting, and ftoring up fo neceffary
" a commodity, that fo an abundance thereof
" may be ready for all fuch fleets and fhips of
" ours, as, fhall be employed that way; the
" which is to be taken off from the colonies at

7   " reafonable

" reafonable and honeft prices, with fuch needful
" merchantdize, as is requirable for their ufe;
" that fo it may become their ftaple commodity;
" and a furer means of fubfiftence, than tobacco
" is likely to be.—And as for the iflands of Ber-
" mudaes; or Summer Iflands, give me leave to
" affure your Lordfhip (as one that well knoweth
" them, and fhall be ready to demonftrate it
" evidently, whenfoever I fhall be called unto it)
" that (in regard of their natural ftrength, the
" fafety of their harbours, their moft opportune
" fituation, their falubrity, and their wonderful
" production) they are the moft advantageous
" piece, not only within his Majefty's dominions,
" but of all thofe parts, for to make ufe of, in
" all thofe Weftern fervices, efpecially fea-em-
" ployments upon thefe coafts; and in that re-
" gard, do well deferve both to be cherifhed and
" well looked into."

It is a little remarkable that this laft advice of
the honeft Captain fhould never have been attended to fooner.   A furvey has at laft been made,
and a harbour difcovered fit for a large fleet: it is
of the more confequence at this time, as thefe
iflands are fituated in the very fpot that could be
wifhed, to annoy the homeward-bound trade of the
French and Spaniards.   It will alfo be an *afylum*
for fickly fhips from the Weft Indies: particular
orders ought therefore to be given, when ficknefs

   is

is likely to become general among a fhip's company, that they depart quickly for Bermudas. On this account, I would recommend an hofpital to be fitted there, with every convenience, on a large fcale, for the ufe of both navy and army: here the convalefcents from the Weft India difeafes might be recruited at a fmall expence, and feafoned to the climate.

From fome recent accounts, I am forry to learn, that among medical gentlemen in the Weft Indies, at this time, there is ftill a confiderable diverfity of opinion; both on the nature of the Fever, and its treatment. Some ftill perfift in the ftimulant plan; after all the evidence of thofe great phyficians which we have quoted, has been brought againft it. This is adding horror to death. We have appealed accordingly to firft principles, and the little experience which we have feen of the difeafe, juftifies the whole. Whoever attempts to controvert the practice of Dr. Rufh, muft bring unequivocal arguments indeed, to fupport his method of cure; for he has to oppofe a phyfician, whofe unwearied zeal and attention carried him beyond the common duties of the profeffion, and have rendered his precepts and authority very high. It appears to me, that the large dofes of mercury act, by their purgative quality, caufing a large and fudden depletion of the inteftines, and the excretories that empty themfelves there. This

medicine

medicine has therefore its peculiar period of exhibition, which ought to be early; and in full
habits, large bleedings ought to accompany it.
Authors, I think, treat the yellownefs of the fkin
too flightly in the pathology of this difeafe. I
cannot fuppofe fuch a large fecretion of bile to
take place, without fome peculiar affection of the
fecreting organ. The large fuffufion of bile
happens particularly during the ftage of debility;
at leaft the fymptoms of increafed excitement
are evidently on the decline. Now, is the liver
itfelf affected? or the vifera connected with the
fyftem of vena portarum veffels, which furnifh
blood for the biliary fecretion? but perhaps the
blood itfelf, during the violent inflammatory ftage,
is altered in its quality, by which means it fur-
nifhes a larger proportion than in its healthful
ftate. This kind of icterus differs from all others,
for there is no obftruction in the ducts: it paffes
into the inteftines, and is taken up by the lacteals;
not the whole of it, for the fæces are more highly
tinged than ufual. In jaundice, from other caufes,
no bile is found in the alimentary canal.

Dec. 3d. Moft of the fhips which have returned this feafon from the Weft Indies, have been
fufferers from the Yellow Fever: yet the difeafe,
in all of them, uniformly difappeared as they increafed their latitude: at 32° north, no frefh

attacks

attacks were known. We muſt ſuppoſe, that communication with the ſhore produces this fever; for a cruize ſeldom fails to put an end to it. Vice-Admiral C. Thompſon, who lately commanded there, was ſo certified of this, from long experience and attention to the ſubject, that he ſometimes inſiſted upon ſhips putting to ſea, when their condition otherwiſe was unfavourable. One caſe in particular, during his command, is too pointed to be over-looked: the Yellow Fever broke out in a ſhip, and her Captain came to inform the Admiral of his diſtreſs: peremptor orders were inſtantly given to put to ſea, without delay. The Captain even heſitated: but the Admiral added, " that if he did not ſee the veſſel under weigh in an hour, the boats of the Fleet ſhould *tow* her to ſea." The ſhip went to ſea; and returned, after a fortnight's cruize, in perfect health; and free from the Yellow Fever, which ſhe carried out.—The Dædelus frigate, that arrived at Portſmouth in October, left the iſlands with this Fever on board; ſo large a number of men and officers were affected, that Captain Counteſs thought it expedient to puſh for Halifax, to land his ſick: but, before he reached that port, the diſeaſe had taken a favourable turn, and was ſoon extirpated.—I apprehend, an imitation of this practice will always be attended with ſalutary conſequences.

Dr.

Dr. Pattifon, who went to St. Domingo furgeon of the Leviathan, has obliged me with the outlines of his practice, when this Fever appeared on board. On the firft fymptoms of attack, he bled the patient two or three times in the fpace of a few hours, if one large bleeding did not give effectual relief. He next gave a large dofe of natron vitriol. much diluted, by way of emptying the inteftines: if this medicine was vomited up, a glyfter of the fame falts anfwered the purpofe; and it was repeated every five or fix hours. It was obferved, that this purge often fat eafy on the ftomach, when other articles, lefs naufeous in other conditions, were immediately rejected. The next procefs was, to wrap the patient in a flannel fhirt, *dipped in vinegar*; which always relieved the comatofe ftate, and the appearance and feel of the fkin were quickly altered. This treatment was fo fuccefsful, that the difeafe became manageable; and a few days commonly reftored the patient to duty*.

Mr. Robert Harris, furgeon of the Thunderer, who has juft returned from Martinique, found a method of cure, much like that of Dr. Pattifon,

* About the time that the Leviathan croffed the line, Dr. Pattifon, at Captain Duckworth's requeft, recommended every man to take a purge of falt water: this was repeated three times before the fhip made the Ifland of St. Domingo; and the good effects of the practice were generally remarked.

equally

equally fuccefsful. He had recourfe to glyfters
of common falt water, in the firft inftance, which,
with plentiful bleeding, were always effectual in
relieving the firft ftage. Mr. Harris thought that
he obtained uncommon benefit, by wafhing the
body all over with the juice of lemons.

The fuccefs which attended the practice of
thefe gentlemen, is a demonftration of the juftnefs
of the principles laid down by Dr. Mofeley and
Dr. Rufh ; and it gives us caufe to hope, that
as they become better underftood, and more generally practifed, the mortality occafioned by this
Fever, may be, in a great meafure, checked. The
relief obtained by the flannel fhirt, dipped in vinegar, and wafhing the fkin with lemon juice, I am
apt to believe, was owing to the cold on the furface produced by the evaporation of the fluid.
I fhould be glad to know the further refult of
this plan, as it muft act powerfully in diminifhing
the excitement, fomething like a cold atmofphere.

The Thetis and Huffar, off the Capes of Virginia, on the 16th of May 1795, engaged five
French armed fhips, two of which were taken.
Thefe fhips had left Guadeloupe lately, and had
buried numbers in the Yellow Fever. From one
of them, the Huffar received the infection, and
afterwards landed eighty-three, ill of the difeafe, at
Halifax. Mr. Ifaac Wilfon, who has favoured
me with this account, attended them for five
weeks

weeks himfelf, in tents fitted for that purpofe: of
this number, ftrange to be told! not one died.

Mr. Wilfon bled freely, in proportion to the
urgency of the fymptoms, and uniformly within
the firft hour from the acceffion; by which timely evacuation, delirium was either completely prevented, or cured. Tartarized antimony was next
exhibited; and care was taken, that no warm water fhould affift the vomiting; which was the
means of relieving the ftomach from vaft loads of
bile, and at the fame time generally procured
evacuation downwards. The bowels were kept
foluble by glyfters or kali tart. A blifter was applied to the region of the ftomach; the head was
fhaved, and cold applications were found ufeful.
The cold bath, in cafes of delirium, was always of
fervice: the fhower bath was the mode practifed;
after which, the patient was wiped dry and put to
bed; fleep followed, and delirium feldom recurred. Such was the practice of Mr. Wilfon, in
eighty-three fuccefsful cafes; a larger proportion
than has ufually fallen to one medical attendant!

CATARRH.

IT is here intended to fpeak only of the Epidemic Catarrh, or Influenza, as it was firft named by the Italians.

This difeafe has twice, in our remembrance, been epidemic in the Channel Fleet ; at one time it prevailed to fuch a degree, in the fpring of 1782, as to render the fhips almoft inactive. The difeafe was then general throughout Europe : it fpread from the fhores of the Baltic to Holland and the Low Countries, from thence to England, to France, Germany, Portugal, Spain, and Italy. The fpring of this year was remarkably late, with a long prevalence of cold eafterly winds : the hedges were not full blown in Cornwall before the beginning of June. A fimilar ftate of weather has commonly ufhered in this univerfal malady. When I was furgeon of the Buftler floop of war, and attending a furvey of feamen at Plymouth Hofpital, I was attacked with the Catarrh. It did

did not appear to me at that time, that I had been
near any perfon under the difeafe : none of the
gentlemen in the room complained, nor was it
apparent among the patients.   My own, was one
of the moft early cafes in the neighbourhood ;
but in a few days it extended over the whole
country.   The firft difagreeable fymptoms, were
an unufual fulnefs about the forehead and eyes.
The eye-balls feemed protruded, inflamed, painful,
and could not bear the light.   There was at the
fame time a large fuffufion of tears, which gave to
the countenance the appearance of exquifite diftrefs.   The lady where I lodged, apprehended
from my looks, fome uncommon misfortune, and
was with difficulty convinced that I was labouring under a very different complaint.   The fulnefs of the forehead was confined to that fpace occupied by the frontal finufes ; and is no doubt to
be afcribed to an inflammatory affeƈtion of the
membranes which lines thefe cavities : the pain
and fulnefs were alfo felt in the upper part of the
nofe, when a fneezing and a difcharge of mucus
became exceffive.   More or lefs of chillinefs and
fhivering accompanied thefe affeƈtions.   I had a
continual thirft, no appetite, though no naufea,
and a total inaptitude to motion.

There is fomething very particular in the propagation of this difeafe, from its fudden attack.
When the fquadron under Admiral Kempenfelt,
confifting

confifting of eight fail of the line, lay in Torbay,
part of the Channel Fleet, Lord Howe with the
reft being in the North Sea, two hundred men
were feized in one night, on board the Fortitude.
The fignal was made to unmoor or get under
weigh in the morning; but the officers could
not get the men out of their hammocks. It was
in vain that they ufed threats, the people declared that they were unable to move. The
furgeon and mates were fent for, who foon pro-
nounced, that they were labouring under a
violent difeafe. This was communicated to the
Admiral; who doubted the report, and fent three
captains with their furgeons, to examine the ftate
of the Fortitude's fhip's company. The captains found it exactly as related; the fhip was
ordered to Plymouth, where numbers of her crew
were landed at the hofpital. The other fhips
had not been two days at fea, till their fituation
was as bad as that of the Fortitude; fome of
them could fcarcely mufter feamen to take in
fail: the whole returned to port. Mr. M'Nair,
an able furgeon, gave me this account of the
Catarrh, in the Fortitude.

In the mean time, the Fleet in the North Sea
fuffered from the Catarrh in an equal degree; and
was obliged to return to port to recruit the people. Some deaths happened in confequence.

On

On the 17th of January 1795, the weather then very fevere, and the thermometer at 17° of Farenheit, I was ordered to examine the ftate of the Cumberland, juft arrived from the Nore, where fhe had been lately commiffioned. I found one hundred and twenty men ill of Catarrh. The fymptoms, however, were in general flight, with no oppreffion on the breaft, and none were confined to bed : a few of the worft were fent to the hofpital. As the acquifition of a 74 gun fhip was of fome importance at this time, the enemy's fleet being at fea, I recommended what was deemed needful in fuch a condition, and reported the fhip fit for fea. Captain Rowley was active and careful in making the fick comfortable. He procured the whole, additional warm cloathing, particularly flannel jackets, which were the more valuable, as a great part of the crew had lately returned from the Weft Indies in the Vengeance. Care was alfo taken not to expofe them to unneceffary cold and wet weather. Directions were immediately given to abftain from wafhing, to fubftitute fcraping and rubbing with dry fand : additional fires were lighted on the lower deck. The fick lift decreafed from this time, and the Cumberland did well.

A few days after my vifit to the Cumberland, I was ordered to report to the Admiral the ftate of the Coloffus, from the number of

fick

fick, upwards of feventy, being returned in the
weekly account. This difeafe was a Catarrh;
and fome were fent on fhore with fymptoms of
inflammation, confiderably greater than any in
the other fhip. The Coloffus appeared to me to
have her ficknefs much aggravated by wafhing
decks; although I found that Captain Jenkins,
then her firft lieutenant, had oppofed the practice, and reprefented it as hurtful. It was afterwards laid afide. The Coloffus did not fuffer
from any future increafe of the fick-lift, during
the cruize*.

The

* " The prevalence of catarrhal complaints appear to me
to have arifen from the intenfe cold. A patient died who was
old, but had been always fubject to complaints of the breaft,
and difficulty of breathing.

(Signed)

Coloffus, Jan. 1795.   J. BALLENTYNE, Surgeon."

" There was a catarrhal fever prevailing towards the end
of the month.   (Signed)

London, Jan.   J. SMITH, Surgeon."

" Catarrh was the prevailing complaint, particularly during
the firft fortnight of this month; having had feldom fewer
than feventy on the fick lift. But on its being fuggefted to
me, that it was much aggravated by too frequent wafhing of
decks; this moft pernicious practice, in cold and damp weather, was intirely laid afide. In its ftead, the decks were kept
clean, by fcraping and fcrubbing with fand, &c. The good

effects

The Fleet being obliged to put into Torbay; in the beginning of February, we there experienced fome fevere weather. The Catarrh was now general in every fhip. Some bad cafes appeared

---

effects of this were foon apparent; for, during the laft ten days, the number of catarrhs never exceeded twenty.
(Signed)
Barfleur, Jan.   P. SMITH, Surgeon."

" The prevailing complaints, flight fevers, and catarrhs, feemed occafioned by the feverity of the feafon: and, no doubt, affifted by that extreme thoughtleffnefs fo peculiar to feamen.   (Signed)
Valiant, Jan.   A. THOMPSON, Surgeon."

" The men have been generally afflicted with violent colds, coughs, attended with hoarfenefs, and fome feverifh heat.   (Signed)
Marlborough, Jan.   T. ROMNEY, Surgeon."

" The catarrhal complaints feem to have arifen from the fudden change of weather; and readily gave way to diaphoretics, opiates, and warm diluent drinks.
(Signed)
Royal Sovereign, Jan.   R. FORREST, Surgeon."

" The prevailing difeafes for this month, have been chiefly catarrhal complaints, attended generally with more or lefs fever; but which, however, have in no inftance proved fatal, or even of long duration. The fever cafes fent to the hofpital, confift entirely of marines, whofe expofure as centinels on the
gang-

peared in the Brunfwick, Canada, and Prince of
Wales.   In the latter fhip it affumed very much
                                                    the

---

gang-ways, in night-watches, to the intenfe colds, which have
lately prevailed, will I apprehend, fufficiently account for
their being attacked, when feamen who keep no watch in
harbour have efcaped.

(Signed)

Aquilon, Jan.   THOMAS MOFFAT, Surgeon."

" About the beginning of January, our people began to
complain of catarrhs ; with head-achs, fome with fore throats,
others with fymptoms of fever.   Thefe fymptoms in general
gave way to the ufual evacuations, and left coughs and pains
of the breaft, which yielded to pectorals.—As far as I am able
to judge, this difeafe did not proceed from any contagion, but
rather from the intenfe cold and wet weather we experienced.

(Signed)

Hebe, January.   JOHN LEGGAT, Surgeon."

" The principal complaints on board the Queen, for the
laft two months, were catarrhs, with coughs, fore throats, and
fevers.   The fevers were ardent, but happily foon fubdued by
the ufual evacuations, bleeding, and laxatives.   The crew at
prefent is healthy, *not* a man in the fhip being confined to
bed, by any difeafe!

(Signed)

Queen, Feb. 28.   ALEXANDER BROWNE, Surgeon."

So completely was the duty in this fhip, adapted to health
by her officers, that Mr. Browne, her furgeon, ufed to fay, that
they left him nothing to do   It was like her method of fight-
ing, *perfect.*   *T. T.*

                                                    " The

the type of a pure Typhus; with weak frequent pulfe, dejection of countenance, great mufcular debility, and ftupor. Among the foldiers who acted as marines, this was particularly remarked.
    They

" The catarrhal complaints, which have been very numerous, in general yielded foon; but fome left very confiderable weaknefs behind: One cafe, a fudden and violent affection of the lungs, proved fatal.

(Signed)

Impregnable, Feb. W. WALLIS, Surgeon."

═══

" The above fevers were, nine fynocha, and one typhus. (Cafes of Fever, in the Monthly Report.) The latter cafe from the beginning had the moft unpromifing fymptoms, and terminated fatally on the 24th day.—Catarrhal complaints are ftill prevailing, but in a milder degree, which I attribute entirely to the abolifhing the wafhing of decks, and the cruize to the fouthward.

(Signed)

Barfleur, Feb. P. SMITH, Surgeon."

═══

" With refpect to the probable caufe, of fo many Catarrhs, with fome fevers, I principally attribute it to the weather being in a very unfettled ftate; for the greater part of them were taken ill the firft ten days of the month, and the weather at that time was very changeable.—Were failors obliged to wear flannel fhirts in the winter months, when they are necefariiy expofed fo much to the inclemencies of a channel ftation, I think it might be attended with good effects *.

(Signed)

Robuft, Feb. J. TURKINGTON, Surgeon."

---

* Cloth trowfers and flannel waiftcoats, have, by recommendations from the Fleet, become a part of Navy Slops. T. T.

B b " The

They had been lately embarked, were very badly cloathed, and, on the whole, looked more like objects for the Infirmaries of the country, than to fight her battles. For these reasons, nine of the

---

" The principal disease has been catarrhal complaints, owing to the intense rigour of the weather; which have given way to gentle diaphoretics and pectorals.

(Signed)

Cæsar, Feb.   J. NUTT, Surgeon."

" A kind of influenza made its appearance on board early in February, but soon disappeared. The medicines employed, were chiefly antimony and nitre; some required bleeding: otherwise the ship's company is healthy.

(Signed)

Niger, Feb.   ROBERT KIRKWOOD, Surgeon."

" The catarrhal affections, were treated successfully by bleeding, antimonials, and pectorals. In the unsuccessful case, there was much diseased secretion from the lungs; blisters, and occasional opiates, were had recourse to, ineffectually. In the other cases, nothing occurred new, or particularly interesting.   (Signed)

Bellerophon, Feb.   R. NEWBERRY, Surgeon."

" The Influenza was the most prevailing disease in this month; it first made its appearance on board the Hannibal, about the 5th or 6th; soon after the commencement of the thaw, during which time the wind blew from the eastward of south: and it continued to affect the officers and ship's company

the worft were ordered to be fent on board the Charon; when wine, æther, and a nourifhing diet, foon brought them about.

In

---

pany until the 12th, when it began to abate, and the wind came to the weftward of fouth-weft.

" It appeared to me, that this Epidemical Difeafe was caufed from a particular ftate of the air, after the thaw took place *.   (Signed)

Hannibal, Feb.   W. WALKER, Surgeon."

* It prevailed in the Cumberland and Coloffus, during the fevereft froft.   *T. T.*

" A fever made its appearance, about the 1ft of February, of the low nervous kind: the fymptoms of which were, naufea, thirft, anxiety, flufhing of the eyes, a pain at the *fcro-biculis cordis*, with a quick pulfe and delirium in the earlieft ftages of the difeafe. Clearing the *primæ viæ* by emetics, in the firft inftance, with blifters, bark and wine afterwards ad-miniftered, were the only fuccefsful remedies employed. From the difeafe affuming an inflammatory appearance, indicated by ftricture on the breaft and full pulfe, with a difficulty of breath-ing, I was induced, in two cafes, to make ufe of the lancet, in both of which it was evidently of differvice. Three died of this fever.

" A fpecies of Influenza, refembling that which made its appearance in 1782, was common throughout the fhip, and I believe the Fleet in general.

                    (Signed)
Melampus, Feb.   J. B. HOUSEAL, Surgeon."

, Thefe two difeafes, probably, only differed in different fub-jects; and acknowledged a common caufe, for Mr. Houfeal does not fufpect any contagion brought to the fhip. *T. T.*

" Catarrhal

In other ships, the inflammatory symptoms ran
high, and venæfection was sometimes repeated
with

" Catarrhal complaints have been moſt prevalent this
month, and in general ſoon yielded to bleeding, antimonials,
laxatives and opiates, and in ſome caſes, bliſters. The
ſhip's company is now almoſt free of it, except one from
irregularity.

(Signed)

La Nymphe.   GEORGE SIBBALD, Surgeon."

===

" The unuſual number who complained on the 1ſt of the
month, I attributed to much irregularity, from a payment to
the ſhip, a few days before the fleet left Spithead, but under-
ſtand that catarrhal complaints with fever, were general in the
Fleet.   There was nothing particular or worthy of obſervation
in the complaints; in two or three days the ſick liſt, from
four and ſix in number, increaſed to twenty and thirty ; and
in a week or ten days, to the uſual number of four or ſix, of
little conſequence.

(Signed)

Santa Margarita, Feb.   J. TOSH, Surgeon."

===

" The feveriſh complaints were, in general, ſlight, and gave
way to evacuants in the courſe of three or four days.   A longer
continuance was attended with ſymptoms of typhus, confined
to weak debilitated conſtitutions, and invariably diſappeared
on the uſe of bark, wine, &c.—The cauſe was, viſibly, intemperance whilſt in harbour ; the ſhip's company having a
quantity of prize-money, and the complaints appearing on
their return from leave.

(Signed)

Pallas, Feb.   R. HARRISON, Surgeon."

" The

with advantage. I vifited a cafe on board the
Canada, with Mr. M'Curdy, with every fymptom
of phrenitis; it proved fatal. But what were the
moft alarming fymptoms in this epidemic, were,
the determination to the breaft, oppreffion about
the præcordia, difficult expectoration, flufhing of
the countenance, as if the circulation had been
compleatly obftructed through the lungs: if thefe
were not quickly relieved by a large bleeding, the
patient feldom furvived the third day. The pains
and ftrictures acrofs the breaft obtained confiderable eafe from a blifter laid over the fternum.
They were, in fome cafes, fo acute, with a feeble
pulfe, and other figns of debility, that æther and
opium were given with the beft fuccefs.

I was now a fufferer in my turn, and was feized
with Catarrh in a violent degree. My confinement was the more to be regretted for four days,
as it deprived me of feeing the practice of the
furgeons. With other fymptoms more peculiar
to the difeafe, fuch as fever, obtufe pain of the
forehead,

" The catarrh, which was for a few days very troublefome
throughout the fhip, appeared very early in the month, and
feemed to be at its height about the 12th; fince which it gradually declined; yielding without much difficulty to gentle
antimonials and expectorants.

(Signed)

Orion, Feb. J. NEPECHER, Surgeon."

B b 3

forehead, difcharge of mucus from the *membrana fchneideriana*, a painful and acute ftitch, drating from my right fide acrofs the cheft, and preventing a free infpiration, gave me confiderable uneafinefs and alarm.   Finding nothing about me indicating pulmonic inflammation, and having been fubject to this kind of pain, after recovery from a fever at Jamaica, formerly, I was induced to try the external application of vitriolic æther. It removed the complaint in half an hour.   The manner of applying it, was by pouring a little into the palm of the hand, and then laying the hand quickly over the affected part, clofely, fo as to prevent the too fpeedy evaporation.

While the Fleet remained in Torbay, and moft of the fhips afflicted with Catarrh, a number of fheep were, by the Admiral's orders, diftributed for the ufe of the fick.

A peculiar conftitution of the atmofphere, certainly caufed this general Difeafe.   There was, befides the cold, which was intenfe, a thick haze, not like the ufual hoar-froft, during a great part of January.   Its different form, in different fhips, cannot be accounted for: even in the fame fhip, among people of fimilar modes of life, there was need of nice difcernment in the treatment.   The very quick paffage to Cape Finifterre, contributed not a little to the health of the Fleet.   The temperature of the air, on the day we made the land,

9

was

was mild as a fummer evening. The catarrhs in fome fhips difappeared almoft at once. This fine weather was followed by a ftrong wefterly wind, which brought us quickly to Spithead.

When a difeafe of this kind occurs in a fleet, the preceding fervice, and condition of the people on board, will very much influence the proper medical treatment to be adopted. Thus, if a fleet, or fhip, had previoufly been a long time at fea, we would not think of carrying the evacuating method of cure, fuch as bleeding and purging, to the length that might be warrantable under other circumftances. When a failor, in harbour, has lived long on frefh provifions, we know that he will be able to bear venæfection, as well as perfons on fhore: but a fea diet does not afford fuch ftimulant nourifhment, and the lofs of blood may be hurtful. Wine was, on that account, very judicioufly prefcribed by fome of our furgeons, and with it, opium, æther, &c. in confiderable dofes.

The general practice, therefore, has been to bleed, when the inflammation run high, to repeat it if the pulfe rofe after bleeding; to keep the bowels open; to give antimonial powder after venæfection, and blifters when pain of the breaft and difficult refpiration were urgent. Thefe, with diluent drink, conftituted the treatment of the Inflammatory Catarrh. When the fymptoms

appeared

appeared more like the low fever, a more ſtimu-
lant plan was practiſed. The fœtid gums were
ſometimes uſeful; wine, bark, æther, and opium,
were given freely. Dyſpnœa was ſuccefsfully relieved by bliſters, and the bowels were kept ſoluble
by very gentle laxatives.

Phthiſical habits bore the attack of this diſeaſe
worſe than others, and their recovery was always
tedious. I would recommend flannel cloathing
to perſons of this diſpoſition, to be worn next the
ſkin; exerciſe on horſeback, when it can be conveniently perſiſted in, is the famous reſtorer of
catarrhal debility.

In the ſpace of five weeks, which comprehend-
ed our abſence from Spithead, from twenty-eight
to thirty died of our Epidemic; a ſmall proportion
out of 28,000 men. Vide the General Abſtract.

MEDICINA

DYSENTERY.

THIS Difeafe has frequently been attended with great mortality in King's fhips, and particularly in tropical climates. It has however, been little known in the Channel Fleet. A few cafes have appeared in fingle fhips, now and then; and, excepting in fome of thofe under the command of Rear-Admiral Harvey, at Quiberon, in the late feafon, the Dyfentery had no fatal tendency. We received a few chronic Dyfenteries on board the Charon; men, juft returned from abroad, where they had the difeafe for months together.

It has been long obferved, that Seamen are peculiarly liable to this complaint, after a long train of ftormy or rainy weather, and when the fhip had been long at fea. Hence it often appears at the fame time with Scurvy, and they are frequently affociated in the fame perfon. But the Flux, in thefe cafes, difappears with the other difeafe, and readily yields to a frefh meat diet, the

citric

citric acid and efculent vegetables; without re-
quiring any of thofe remedies, more particularly
adapted to other fituations.

We have alfo known the Flux combined with
Typhus, as in the tranfports with Lord Moira's
- army: on the coaft of Africa, and in the Weft
Indies, I have feen it joined with intermittent and
remittent fevers.

After the memorable hurricane, on the 5th of
October 1780, the Dyfentery broke out in the
Berwick, then in the fquadron under the command of Rear-Admiral Sir Joſhua Rowley. This
fquadron had accompanied the homeward-bound
Jamaica Fleet to the coaft of America, and was
returning to Port Royal, when overtaken by this
tremendous ftorm.   Owing to the hatchways not
being fufficiently fecured, an immenfe quantity of
water was let into the hold, and over the decks,
ſo as to endanger the ſhip.   After the mafts and
bowfprits were thrown overboard, the motion of
the ſhip exceeded all defcription : to prevent the
ſhot and other things from rolling about, or hurt-
ing the limbs of the people, while employed at the
pumps, the bags and hammocks were cut down.
By doing this, the bags, bedding, and cloathes of
the people, were foaked in water ; and being but
imperfectly dried before they were flept in, did
not fail to injure the health of the ſhip's company.   Many of them flept for nights together

on

on the wet decks, overcome with fatigue, and
debilitated from the want of food ; for no provifion
could be dreffed in the boilers for fome days. We
had now been eight weeks from Port Royal; when
Captain Keith Stewart, the fenior officer of the
fhips that remained in fight on the fifth day
after the gale, thought proper to fhape his courfe
for England in the Berwick ; having ordered
the others to make the beft of their way to Jamaica.

We had fcarcely got our jury-mafts rigged, and
fail fet, when the Flux began to make its appearance. It fpread with great rapidity, and thirty of
the beft men on board died from it, or the Scurvy,
before we made the Englifh land, which was
feven weeks from the time of the hurricane. During this time, the people were on a fhort allowance of water, as well as of every kind of provifion. The hold was filled with the fick ; and the
groans of the dying, or thofe in acute pain, were
heard on the quarter-deck. Some of them died
in a furious delirium, uttering the moft blafphe-
mous oaths and horrid imprecations that could
be conceived. This is a fpecies of raving in difcafe, that I have only met with among the de-
bafed, moft ignorant, and uncultivated minds.
What added much to our afflictions, was a mif-
fortune that befell the medicine cheft, during the
gale, the greater part of its contents being deftroyed,

ftroyed ; and by which means little was faved for
the cure of this fatal difeafe.

The fituation of the officers in point of cloathing, was little better than that of the feamen.
Our beds were foaked in water, that paffed
through the feams of the decks ; and every trunk,
cheft, and bureau, were dafhed in pieces.   After
being for fome nights without fleep, and at laft
overcome with fatigue, I ftretched myfelf upon a
wet fail, where I lay found for fome hours.   It
was not long before I was attacked in my turn,
and had reafon to think my complaint was
brought on from fleeping on the fail.   A midfhipman who lived in the fame mefs, was feized
at the fame time.   We complained of violent
griping, a coppery tafte in the mouth, with ficknefs, and inclination to vomit : the tafte of copper was fo particular, that we could not help thinking it muft have been taken in pea-foup, on
which we dined, that came from the boilers.
The cook, however, affured us, that could not
happen, as they had been uncommonly well cleaned that morning.   There were alfo fymptoms of
fever, which ufhered in this painful difeafe, and
there was more or lefs of a cold and hot fit in
the beginning.   Towards evening, we took emetics.   The midfhipman had a full vomiting, and
felt little from the complaint afterwards : in my
cafe, the emetic fcarcely excited vomiting, and
from

from being in much pain, had not the refolution to encreafe the dofe; but to this neglect had I to date a long confinement, which reduced me to the laft ftage of emaciation and debility.   In the mouth of the Channel, we fell in with two Englifh frigates, from one of which we received a little opium.   No length of time will ever efface from my memory the relief which I obtained from the firft dofe : fo quickly did it operate, that I feemed at once to have recovered perfect health. From this time, I was relieved from the moft painful fymptoms, the *tormina* and *tenefmus*, by repeating the opiate, and it was full two months before I could leave it off altogether.

There was little doubt but this Dyfentery was caufed by the beds and cloathing of the feamen being foaked in water during the hurricane ; the cold weather, fatigue, and falt provifions, might alfo have their fhare in it. Seldom has a king's fhip arrived at Spithead in a more diftreffed condition than the Berwick at that time : officers and men with little cloathing, emaciated, and pale, having been fifteen weeks from port ; we were compared to fo many dead bodies, that had been under water for a week, and then reanimated. Near three hundred of the fhip's company were ill ; fome of the fcorbutics, worn out, died by moving into the boats, and many others died afterwards at the hofpitals.

There

There is a very painful fymptom attends this difeafe ; and I am furprifed to fee it doubted by one of the beft writers on the fubject. It is the ftrangury, which fometimes terminates in a total fuppreffion of urine. It would appear, from the almoft conftant tenefmus, that the rectum is pro- truded confiderably ; this portion of it being con- nected to the fundus of the bladder by cellular fubftance, bears it in fome degree downwards, by which means, the mouth of that vifcus is me- chanically elevated, and moved from the direct entrance of the urethra ; and thus the Dyfuria Dyfenterica is produced. Whether this explana- tion will fatisfy other people, I cannot tell ; but it appears fo to me.

A medical man, like other adventurers, fome- times acquires profeffional knowledge from perfo- nal fufferings : the danger which I had efcaped, and the diftrefs that I underwent, made me re- folve to pay particular attention to this difeafe in future. Since that period, I have treated it, in a variety of cafes, on the coaft of Africa, in the Weft Indies, as well as in England, among fea- men, and in moft other conditions of life.

I believe it will happen more frequently elfe- where, than among failors on board, that blood- letting may be found neceffary in a beginning Dyfentery ; there are certainly, in fome cafes, evident fymptoms of encreafed excitement ; and probably, a tendency to local inflammation in
fome

some part of the inteſtinal canal. Where the ſtomach is much affected with bad taſte, nauſea, inclination to vomit, I think an emetic adminiſtered, to excite a free diſcharge of its contents, ſhould never be omitted in the beginning: if the medicine given can be ſo managed, as to act as a purgative at the ſame time, ſo much the better. For this purpoſe, twelve or fifteen grains of ipecacuanha, with a grain and a half of tartarized antimony, divided into three doſes, and given at the diſtance of half an hour, will generally anſwer the intention; otherwiſe a ſaline purge ſhould be taken ſhortly after the emetic. This being done, unleſs ſymptoms of increaſed excitement prevent it, I always give an opiate combined with ſome antimonial preparation, with a view to open the ſkin. The patient ought to be confined to bed during its operation, which ought to be encouraged by drinking plentifully of ſome grateful liquor..

According to the urgency of pain, tormina, and teneſmus, I repeat the purgative medicine; and always follow its operation with the opiate and antimonial, or opium with ammonia. The reaſons for condemning free doſes of opium in this ſtage of Dyſentery, ſeem to have been owing to the neglect of giving a purgative before. If, therefore, the purge has not preceded the opium, the latter ſeldom fails to do harm. The griping that is commonly aggravated after the ſpace of

ſix

fix or eight hours from the taking of the opiate, can only be relieved by a brisk purge : but when the one follows the other as we have directed, the disease has rarely continued long or obstinate. In the latter stage, where astringents have been recommended, unless their exhibition has been preceded by purging, they have generally done harm in my hands. The neutral salts well diluted, I have always found the best purges *·

When the Dysentery has advanced to its chronic stage, as happened with myself, it appears to be continued more from habit than other causes. In my own case, whenever the opium was intermitted

* "In a former situation, I had many opportunities of seeing Fluxes in their worst stages; which naturally induced me to pay particular attention to that disease; consequently I took some trouble in giving the most of doctrines, I conceived at all rational, a very fair trial. I beg leave to say, I am decidedly of opinion with youself, respecting the opium being given in large doses, after a few gentle purges of sal. cathart. * You will perceive by the report above, I have had seven Dysenteries in the month ; four of whom are recovered, and the remaining three in a very fair way. I had formerly one hundred and twenty fluxes, during a cruize of four months, and was fortunate enough not to lose a single patient by the same practice.   (Signed)

Orion, May 1794.   WILLIAM PATTISON, Surgeon."

———

* Mr. Pattison here alludes to a paragraph in the second edition of my Observations on Scurvy.

ted at bed-time, although the dose never exceeded five grains, the liquid dejections returned, with more or less of griping at the same time. In this manner it continued with me for two months, during which I acquired flesh and strength, that at last allowed me to diminish the medicine gradually, and leave it off. Much, however, might depend on my constitution, and the period of life, for I was then under twenty. The cure, therefore, seemed to be effected by the opium overcoming the habit; which enabled the natural actions of the intestinal tube, to recover their former propensities.

The vitriolic solution, so strongly recommended by Dr. Moseley, has always failed in my practice; nor have I found any reason for preferring the mercurial treatment of Doctors Houlston and Clarke. Riding on horseback, with other active exercises, and flannel cloathing, have done more in the chronic Dysentery than any remedy which I have seen tried; but the early stage of the disease is the fittest period for the cure, and if omitted, pain and suffering are the lot of the patient.

I have only dissected one subject for this disease; nothing uncommon was detected: nor do I think that the dissections of other physicians, have thrown any light on the pathology of Dysentery, so as to direct a better method of cure.

———————

C c

MEDICINA NAUTICA.

SMALL POX.

IT appears, from the Medical Regifter of the Fleet, that the Small Pox has been a very frequent difeafe in the fhips. There are many reafons for this. The frequent communication with the fhore, and the conftant vifiting between the wives of feamen and their hufbands, are generally the means of conveying the infeftion on board : we have even feen an inftance in the Orion, of a man bringing the difeafe from the ward at Haflar. Thefe are fufficient excufes for me taking notice of it, in a work on the Difeafes of Seamen.

Seamen brought up in our merchant fervice, go to fea, in general, very young. Many of them have efcaped Small Pox, and while they continue on long voyages, they are exempted from all intercourfe with infefted perfons; in this manner they grow up to manhood; and, when brought on board his Majefty's fhips, are expofed to the difeafe, as we have feen. During a war, the
Small

Small Pox ward in Haſlar Hoſpital is ſeldom
empty. This circumſtance gives great force to
the arguments of thoſe enlightened and charitable
members of ſociety, who are advocates for a general inoculation. What a pity it is, that ſo
valuable a part of the community as our ſeamen,
ſhould be ſubject to the ravages of ſo mortal a
diſeaſe in their adult age. A very large proportion, in every King's ſhip, are men of this deſcription: in a ſhip of 74 guns, I have known
ſeventy ſeamen, that never had been conſcious of
going through the diſtemper.

Whether ſome of the evils attending Small
Pox, as now narrated, might be remedied, remains for future experience to determine. It appears to me, that a general inoculation in a ſhip,
or fleet, is highly practicable. I have not been
informed, that any thing of the kind was ever
done in a man of war, before practiſed under my
directions in the Charon and Orion; but the
ſucceſs which attended the meaſure, in both
theſe ſhips, juſtifies the principle.

A general inoculation might be done while a
Fleet is in port, by ſelecting the ſubjects into an
hoſpital ſhip, or let a pavilion of Haſlar Hoſpital
be allotted for the purpoſe. A week, or ten days,
of ſuitable diet and regimen, are all that is neceſſary for preparation. In order to reconcile the
ſeamen to this practice, it would be neceſſary to

circulate

circulate among them, in printed papers, the various arguments and information which point out the fafety and utility of inoculation. They might be told, that they are in conftant danger of catching the infection, whenever they go on fhore; and the longer they delay inoculation, the hazard becomes greater. If any religious fcruples fhould appear, as happened in the Orion, they ought to be combated by the chaplains of the fhip, in the manner that has been ufed elfewhere by other clergymen. There is a valuable little work on this fubject, by the Rev. Mr. Turner, of Newcaftle upon Tyne, written in a familiar ftyle, and intelligible to every capacity.

When the Small Pox appear in a fhip at fea, where there is an hofpital fhip belonging to the Fleet, the patient ought to be immediately moved, and with him, every thing that is likely to imbibe infection. When fhips are cruizing, it will be neceffary to keep the patient at a diftance from the fhip's company. I have heard of Guineamen towing them in a long-boat. The boats are the beft places in this fituation, on the booms, or flung at the ftern : centries may then be placed to prevent communication with the fhip's company. If there are any perfons on board who have never had the difeafe, I would recommend inoculation as foon as matter can be taken from the difeafed fubject. A captain and

officers

officers might, by attending to this bufinefs, meet with fome additional trouble; but it would be amply repaid, by alleviating one of the moft fatal diftempers that human nature can undergo.

With regard to any preparatory courfe of medicine for the Small Pox, I have only to remark, that I have but little faith in its effects of producing a mild difeafe. In my opinion, the artificial method of ingrafting the poifon, is the fole caufe of the diforder going through a moderate degree, both of fever, and eruption of puftules. Whether this depends upon the fmall quantity of variolous matter introduced, or other adventitious circumftances, I will not decide. I think, however, when we can mend the habit of body, it ought always to be attempted. Nature, it would feem, is taken by furprize, by inoculation, and only difpofes a fmall part of the fkin to propogate the variolous matter, as if fhe did it with a view to get eafily fecured againft a dangerous diftemper *.

* Strong mercurial purges, in full habits, are famous, during the eruptive fever: I fufpect that they act chiefly by the fudden depletion which they occafion, by evacuating the contents of the bowels.

R̶ Calomel - - - g^r vij.
  Pulv. Jalap. - - g^r xxv.
  Syr. - - - - f^t bol.
   mane fumendus.

In

In one feafon I inoculated near three hundred
children, all of whom came fortunately through.
There was a great number of them under five
months; and I think this is the beft age for the
practice, although the habit feems lefs fufceptible
of receiving the difeafe by the common method
of puncture with a lancet.   I never obferved any
other diforder, that could be certainly known to
have been inoculated along with the Small Pox
pus.   This difeafe often brings to action, complaints that have not been noticed in the body
before; particularly fcrophula, and cutaneous
affections.   It was a very common thing for
parents to fay, that their children received fuch
and fuch difeafes by this means; but I knew that
they were latent in the conftitution before.

RHEUMATISM,

MAY juftly be claffed among the difeafes of feamen. Their laborious duty, expofure to weather and climate, and the accidents to which they are continually liable, make this a very frequent complaint in our fhips. Hence it is often feigned for the purpofe of effecting their difcharge; and one of the chief fources of their tricks and deceptions. When this is the intention, the ftory is commonly introduced by the account of fome old hurt or fall, though it may have happened years before, in order to obtain a more patient hearing from the furgeon. The Pfeudo-Rheumatifm is, for the moft part, confined to the loins; and to ftrengthen the pretenfion of being objects for invaliding, an incontinence of urine is faid to accompany it. A perfon who has been accuftomed to hear their narratives, may eafily diftinguifh the real, from the affumed diforder: the tale is feldom contrived

with

with addrefs, and it requires but a flight crofs-
examination to detect the incongruity of its materials.   Yet obftinacy from difappointment will
occafionally carry them to great lengths in their
perfeverance; and the fictitious complaint fometimes becomes a fatal difeafe.   The applications
of copper to their fores has killed many.   Thefe
deceptions alfo afford them opportunities for
making their efcape from the hofpitals.   I have
known a man, that for two or three months had
never been known to raife his head and fhoulders
beyond an angle of forty-five degrees, with the
reft of his body; yet, when he found his purpofe
fit for execution, would get over the hofpital-wall,
and make his efcape.   Thefe tricks have, however, been lefs fuccefsfully practifed, and are now
more rarely met with; partly from the inquiries
which have taken place in the hofpital departments, by both furgeons and officers; and partly
from the laudable improvements in diet and neceffaries, with which our fick have been furnifhed
in their own fhips, that bind them by gratitude
to the fervice.

 This difeafe appears in all feafons and climates,
but particularly where viciffitudes of heat and
cold are frequent.   Ships leaving a warm climate,
and returning to England in cold weather, when
the people are thinly cloathed, often fuffer from
Rheu-

Rheumatifm *. A long train of rainy and ftormy
weather, in the Channel, alfo renders the body
more liable to thefe pains: the mufcles of the
joints, near old wounds, bruifes, or broken bones,
are the parts more readily affected with the chronic difeafe. The acute Rheumatifm appears moft
often in thofe conftitutions liable to other inflammatory affections.

At Jamaica, in June 1784, after being ten
months on the coaft of Africa, I was feized with
the Rheumatifm. It obferved a regular perio-
dical form. About an hour after noon it was
ufhered in by a febrile paroxyfm, that continued
more or lefs till midnight. A fwelling and inflammation appeared on the firft day in one leg,
but afterwards in both, over the ancle-joint, and
about the upper part of the foot, attended with
acute pain, which fubfided with the fever. In the
morning and forenoon I remained tolerably free

---

* When a fhip returns from a warm climate to England,
during the cold months, it will be great humanity in the
officers to have a ftock of warm flops in referve; for the feamen are not provident enough to think of this themfelves.
If they have been fome years on a foreign ftation, their
cloathing has either been worn out, or fold, fo that nothing
but attention in their commanders can fave them from fuffering by cold weather.—The duty of boats is another caufe of
frequent Rheumatifm. It is a pity that fome accommodations
for drying their cloathing could not be made at the landing-
places, at every fea-port.

from

from uneafinefs, till near one o'clock, the fever, fwelling, rednefs, and pain returned, and went through a courfe fimilar to the preceding pa-roxyfm. This was my fituation for fome days; when the Rheumatifm difappeared, and the fever affumed the remittent type peculiar to that country. My confinement afterwards was long, and painful; a quick paffage to Europe, in the Auguft following, only faved me from the grave.

This difeafe appears to me to refide more in the mufcular fibres than phyficians have ufually admitted, The acute fpecies is an increafed degree of their excitement, attended with fynocha. The Chronic Rheumatifm, which follows the Acute, is the exhaufted ftate of thefe fibres. Where the Chronic is the primary difeafe, it is always induced by debilitating powers, and generally in parts previoufly weakened, as, after diflocations, broken bones, wounds, or bruifes, Thofe obftinate complaints which caufe permanent lamenefs, may be attributed to a torpid ftate of the mufcular power; or even a paralyfis of the mufcle, which becomes no longer fufceptible of ftimuli; or has loft that difpofition which fitted it to recover its exhaufted fpirit of animation. The pain is felt at the joints, as being the centre of motion.

The cure of the Acute Rheumatifm is like that of all other inflammatory difeafes. Bleeding is, therefore, the principal remedy; and, in proportion to the early time in which it is performed, ed, it will do the more good. It is at the very firſt attack of the fever and pain, that the ſuccefs is moſt certain; a large bleeding, ſuited to the ſtrength of the patient, will ſave much future pain. If that has been neglected for the firſt days of the illneſs, the effects of venæſection become more doubtful; and the danger of converting it into a chronic diſeaſe, ought to be care · fully weighed before it is performed. This being done, every thing beſides that can reduce the powers of the body, fhould be immediately practiſed. Total abſtinence, or confining the patient to water gruel alone, are proper; and the belly muſt be emptied by briſk purges, which include mercury in their compoſition. All external applications at this time are of ambiguous efficacy, and I have ſeen them do much harm.

The cure of Chronic Rheumatiſm has, I believe, been generally conducted on a ſudorific plan. But, however extenſive this practice may have been, the propriety of it appears to me ſomewhat queſtionable. Expoſure to cold after being over-heated, wet cloathes, damp beds, or damp lodgings, a part of the body being expoſed while the reſt is kept warm, &c. are among the cauſes which moſt commonly produce it : a dry ſkin, and diminiſhed perſpiration, are more obſervable than happens in other diſeaſes. Theſe cauſes, joined to the appearance of the ſurface, ſeem to

have

have led phyficians to the adminiftration of fudorific medicines, with a view to reftore the diminiſhed cutaneous diſcharge; and to this, the method of cure has been almoſt confined. If I was
to ſpeak of the ſuccefs of the ſweating procefs,
in my own hands, and what I have ſeen in the
patients of other phyficians, in ſome thouſands
of cafes, I would very much doubt, whether any
of them was relieved by medicine. In ſome of
theſe cafes, the ſweating has been continued for
nights and days, without the ſmalleſt abatement
of pain; but with an evident increaſed degree of
debility, which I alfo confider an increaſed degree
of the difeaſe. Inſtead, therefore, of wearing out
the already exhauſted fenforial power, by difpro-
portionate ſtimuli, I would have the body gradually recruited by nouriſhing diet, gentle exercife, or fuch as the patient can bear, a warm
climate, flannel cloathing next the ſkin, to be frequently changed, and ſtimulating medicines accommodated to the ſtrength of the patient.
When flannel cloathing is wore next to the ſkin
but a few days, it ceaſes to perform many of its
good offices. The perfpirable matter foon foftens
the hairy part of the cloth; for I confider it as a
conſtant fleſh-bruſh to the furface, which ſtimu-
lates the ſkin, and excites its excretories, to fupport an increaſed action: befides, it confines the
heat of the body more than linen or cotton, by

§   which

which means the variations of weather, and a changeable atmofphere, are fuftained with impunity.

The beft cures which ever came under my obfervation, in Chronic Rheumatifm, were the cafes brought in the Charon's hofpital, from Rear-Admiral Harvey's fquadron, in Quiberon-bay, in November 1795. Thefe were feamen much debilitated : but the comfortable and cleanly lodgings of the hofpital fhip, with a diet agreeable, and nourifhing beyond what is ufually met with in the hofpitals on fhore, foon reftored them, with the affiftance of very little medicine *.

With refpect to external applications in the Chronic Rheumatifm, they appear to act by exciting

---

* The Charon's diet for the fick, at fea, partook of the oddities which have marked the character of the Phyfician to the Fleet; it was not modelled to the rules of either a fteward, or purfer, but the appetite of the patient. Frefh mutton broth, with abundance of vegetables, was prepared every day: the meat roafted, or made into a chop, if it was preferred : egg flour pudding, or rice pudding; occafional cuftards; eggs dreffed in different ways; foft bread, baked daily ; pea-foup, nicely feafoned; pan-cakes; apple-dumplings; orange marmalade; pickles of different kinds ; fowens, with wine and fugar, a fupper for convalefcents; tea, coffee, or cocoa; wine, porter, punch, negus, &c. &c. Nor was there ever found an excufe for not having all this at fea.

December 9th. Captain Taylor, of the Fame, affures me, that molaffes mixed with the flour-pudding, is an excellent fubftitute

citing the fkin, and other parts near the feat of pain; and are to be much affifted by the friction ufed in their application. Æther, with the hand clofely laid over the fpot, to prevent evaporation, acts as a gentle ftimulant. Blifters fhould always be removed before the veficles rife: otherwife they act as unnatural ftimuli, by exciting the neighbouring veffels beyond meafure.

Sea-bathing, in the proper feafon, and exercife daily perfifted in, are the beft fupporters of the reftorative regimen, which we have recommended in this capricious and obftinate difeafe. I cannot help lamenting, that a fuit of baths is ftill a defideratum at our hofpitals.

---

fubftitute for eggs: with this addition, I beg leave to call it an
USHANT PUDDING,
from having been highly relifhed in the late cruize.—I muft now requeft, that the Officers of the Fleet will communicate this intelligence to the feamen, as the Lords of Admiralty have determined, that molaffes fhall form a conftant and permanent part of fea-diet. One of our Admirals emphatically calls it the *fafety* of the Britifh Empire.

MEDICINA NAUTICA.

INTOXICATION.

THE frequency of Intoxication among Seamen; the difeafes which it produces; and
the number of deaths which happen in confequence, fufficiently point out this fubject as an
object of medical inquiry. Many fatal cafes have
occurred in the Fleet, and one under my own
eyes in the Charon. There are many circumftances connected with a fea-life, which lead to
the indulgence of inebriating liquors, and often
confirm it into habit. The early entrance on
fhip-board, whether in his Majefty's fervice, or
that of the merchants, before an education has
been compleated to regulate the moral conduct;
the bad example of others; and the abominable
cuftom of grog-drinking, lay the firft foundation
for this moft pernicious practice: to all thefe may
be added, thofe merry-makings and gufts of joy,
which the thoughtlefs failor plunges himfelf into,
when he returns from a long voyage, and with
plenty of money in his pockets. Inconfiderate
beings! after having braved both the battle and
the

the ftorm, how often does the Lethean cup arreft that courage, which feared neither the arm of man, or the war of elements!

This holds up another argument for extending the ufe of malt liquors in the feaman's diet ; and could be eafily done, by the method recommended in a former part of this work. It does not appear, that habits of drunkennefs prevail more in the navy than formerly : I am even informed, from good authority, that feamen, when on fhore on leave, at this time, confine themfelves more to the ufe of ale and porter, than was obferved on former occafions. Now it would be but a fmall attention on our part, to encourage the indulgence of malt liquors ; health and morality claim it from us : there are facts in thefe pages, that point it out in language that cannot be refifted ; and I hope the Board of Admiralty will fee the neceffity for adopting a meafure fo falutary.

It was my acquaintance with a fea-life, and from having known fo many fatal cafes of Intoxication, that induced me to make it the fubject of my Inaugular Differtation at Edinburgh, in 1788 *. Since that time, I was encouraged to profecute the work on a more extended plan ; but before

* " Differtatio Medica Inauguralis : Quædam, De Ebrietate ejufque Effectibus in Corpus Humanum complectens.
Edinburgi, 1788."

before I had compleated my tafk, the war with France broke out, and I was obliged to leave it, without any fixed refolution of following it up at a future time. What I have therefore to offer here, will be a few practical remarks, more immediately connected with my prefent undertaking.

We have often known the bad effects of Drunkennefs, in predifpofing the body to receive contagion; it has been in our late tranfactions particularly marked by Mr. M'Callum of the Valiant, and Mr. Kenning, of the Invincible. In Scurvy, we know it to be equally hurtful, but fingularly fo, when occafioned by fpirituous liquors. In fituations where feamen are expofed to marfh effluvium, Intoxication has been one of the chief caufes for rendering it active in the body: it cannot therefore be too ftrictly forbidden by officers, who wifh to preferve a healthy fhip's company.

The treatment of the Drunken Paroxyfm, or fit of Intoxication, is the moft frequent part of a navy furgeon's duty; and it requires inftant relief. The correction of habitual drunkennefs in fhips, belongs to officers; the example of fobriety in themfelves, the due obfervance of good order and difcipline, with the exemplary punifhment, are the moft effectual means for this purpofe.

D d

When

When there is danger of fuffocation*, as it is called, the patient ought to be placed in a pofture, that may keep the head and fhoulders erect: it ought to be in a place freely ventilated and cool; the neckcloth and collar of the fhirt muft be unbound, and no perfon ought to crowd round the fpot, but thofe who may be wanted as affiftants. It is of fome importance to know the nature of the liquor that has been drank, and the quantity. It will always be proper to provoke vomiting as foon as poffible; and this will be eafily accomplifhed by tickling the throat with a feather, or a finger. If raw fpirit has been taken in great quantity, and the patient is able to fwallow, he may be allowed to drink plentifully of water either warm or cold, and then the tickling of the throat, to urge the vomiting, may be renewed. This ought to be continued, till the whole contents of the ftomach are evacuated.

When there is danger of apoplexy, which is to be diftinguifhed by the bloated countenance and ftertorous breathing, blood-letting becomes indifpenfable. Blood ought therefore to be taken from the temporal artery, the jugular vein, or the arm, from a large orifice, in quantity as much as the perfon can bear. It muft however be limited occafionally, by weighing the ftate of debility

* A flighter degree of apoplexy.

debility that will fooner or later fucceed. At the fame time cold applications ought to be laid on the head, fuch as a cloth wrung out of the coldeft water, and frequently changed. Snow or ice may be applied, if at hand : but the water for this purpofe may be cooled by common falt or nitre, and the cloths dipped in during their folution. It is furprizing to fee how quickly danger is removed, by unloading the fto-mach, and it is by far the moft certain remedy. A failor belonging to the Vengeance, then lying in Prince Rupert's Bay, Dominica, in June 1793, drank near three pints of new rum, and was faved by exciting vomiting; the debility that remained was very great.

When a drunken perfon has flept long, expofed to cold or rain, the treatment will very much depend on the degree of collapfe or debility, that has taken place from the exceffive exhauftion of fenforial power. If the extremities are cold or froft-bit, it will be proper to rub them well with fnow or very cold water, and not to bring him too quickly into a warm room, or near a fire. The exhibition of any thing by way of medicine, muft be done with caution, and in very fmall quantity at firft. It may be then gradually increafed, as ftrength revives; and diet and medicine adminiftered, as cuftomary in fimilar ftates of weaknefs.

I have

I have frequently feen a drunken failor fall overboard, and when picked up, be perfectly fober. I have not had fufficient experience to know how far this might be imitated in practice; but I have a notion, that fprinkling water on the body might be ufeful, as we fee good effects from cold applications to the head.

MEDICINA NAUTICA.

SCURVY.

WHEN I laſt had occaſion to lay my thoughts on this Diſeaſe before the Public, I could not help congratulating the Country and myſelf, on the proſpect of exemption from it, in the Channel Fleet*. Shortly, however, after this work had iſſued from my hands, unforeſeen cauſes began to operate, which in a few weeks produced a more general Scurvy than had been ever known on home ſervice.

The froſt in the month of December 1794, ſat in extremely ſevere; and a cold north-eaſterly wind prevailed for three or four months. About Chriſtmas, the French taking advantage of our fleet going into port, aſſembled, with aſtoniſhing clerity, at Breſt, a fleet of thirty-four ſail of the line, beſides frigates, with which they put to ſea, in order to intercept our outward-bound Weſt India convoy, under Rear-Admiral William

---

* " Medical and Chemical Eſſays. Jordan, London, 1795."

Parker.

Parker.   Many of our ſhips were refitting; ſome
of them in dock, the whole having ſuffered from
a long courſe of boiſterous weather, in September, October, and November preceding.   It
was therefore with great exertions that a Fleet of
thirty-three ſail of the line were able to ſail from
Spithead on the 26th of January.

In the mean time, the enemy's fleet at ſea had
experienced a continued ſtorm, from which it was
reported that five ſhips of the line had foundered,
with their crews, and that the others were much
diſabled.   In this condition they returned to
port.

The Channel Fleet having put into Torbay,
from contrary winds, experienced much cold
weather, and a dangerous gale of wind from the
ſouth-eaſt.   At laſt a northerly wind brought the
large convoy out of Plymouth : Earl Howe having ſeen the outward-bound ſo far to the ſouthward as Cape Finiſterre, returned to Spithead on
the 26th of February.

While the Fleet lay in Torbay, no freſh beef
was ſerved to the people, but mutton for the uſe
of the ſick only; by which means we were full
five weeks on ſalt proviſions, when the firſt freſh
meat was allowed at Spithead. During this time,
an Epidemic Catarrh had raged in every ſhip, and
the debility which followed it, had certainly ſome

3   ſhare

fhare in predifpofing the body for the attack of the Scurvy *.

On our return to port, from the number of cattle that had already perifhed from the rigour of the feafon, beef was much rifen in price, and the contractors were not able to purchafe. The Victualling Board, on this account, thought proper to difpenfe with the ufual allowance to the fhips, and reduced it to one day's frefh meat in the week. Whether there was more œconomy, in point of expence, in giving falt provifions, than frefh, is matter of doubt with me; but this I well know, that fuch a change in the victualling of a large fleet, ought never to have taken place without confulting officers on the fpot.

Independent

* "Catarrhal complaints were the moft prevailing during the month of March and part of April, many of which terminated in Scurvy. During our late cruize, numbers were afflicted with that malady: the citric acid, to the quantity of three ounces per day, cured many, and always ftopped the progrefs of the difeafe. It was given with wine in the following manner:

    R Vin. Rub. - ℥ij.
    Suc. Limon. - ℥i.
    Saach. - - ℨij. M.
   Fᵗ· hauft. ter die fumendus.

Some of the Catarrhal complaints terminated in remitting fever.

     (Signed)

       W. WALKER, Surgeon,"

Hannibal, April 1795.

Independent of the effects which an extreme degree of cold, long continued, may poffefs in predifpofing the body to Scurvy; it had at this time deftroyed all vegetation in the neighbour-hood of Portfmouth. Even the bum-boat wo-men, who are in the practice of bringing vegeta-bles to fell in the fhips, could procure little of the kind, or at a price that put them out of the reach of the failor.

In this ftate of the Fleet, my fears were very early alarmed. I recommended the immediate ftop to be put to the ufe of falt provifions; and early in March laid before the Admiral, what changes of diet I thought would be ufeful under the prefent circumftances. They were tranfmit-ted to the Admiralty; and by their Lordfhips referred to the Victualling Board. The Commif-fioners of Victualling, *difapproved* of the whole, the ufe of molaffes excepted; but a fupply of them did not come in time to our relief.

A fquadron, of five fail of the line, failed on the 17th of March, under Admiral Colpoys; to each of thefe fhips I fent a cafk of lemon juice; at the fame time informing the Admiral of my reafons. This fquadron, after a month's cruize, returned to port with many ill of Scurvy, and a general dif-pofition to the difeafe throughout the fhips: fome very bad cafes were even fent to Haflar,

from

from the Aftrea frigate. (Vide General Abftract of Health.)

A fquadron, of four fail of the line, returned from the North Sea, under Rear-Admiral Harvey. The Prince of Wales, with a raw crew, and a bad party of foldiers, inftead of marines, on board, had fent fifty fcorbutics on fhore to Deal Hofpital; many of thefe in the laft ftage of the difeafe, five of whom perifhed in the boat. The cold of the North Sea is generally reckoned more fevere than in the Channel; hence the Prince of Wales and Ruffel, feem to have had worfe cafes than with us at Spithead.

The Scurvy, towards the beginning of April, had made very confiderable progrefs, particularly in the Excellent, Minotaur, and Invincible. Thefe fhips were part of a fquadron, ordered to fea, under the Honourable Admiral Waldegrave. I had previoufly reprefented the ftate of the Fleet to the Commiffioners of Sick and Wounded; and demanded a fupply of lemon juice, or the fruit in its entire ftate, to be ready on board the hofpital fhip, to ferve the detached fquadrons as they went to fea. This fupply not having come in due time, in confultation with Admiral Waldegrave, and by his order, I purchafed at Portfmouth feventy pounds worth of lemons and oranges, which were diftributed to the fquadron. At the fame time I informed the Sick and Wounded ed Board, that I did not think thefe fhips were in a condition to go to fea, without being provided with articles for the cure of Scurvy. Thefe fhips continued at fea feven weeks, and towards the end of the cruize had a further fupply of fruit, which the Admiral ordered to be purchafed from a Swedifh veffel. The confequence of thefe pre-cautions were, that many cafes were cured at fea, without a fingle death : but the crews of the whole were much in need of frefh meat and other delicacies, when they returned to port.

Admiral Waldegrave was fo obliging as to tranfmit to me, copies of remarks on the fruit, in the treatment of Scurvy by the furgeons of his fhips. It was obferved in each fhip, that by changing the beer for grog, had a quick effect in encreafing the number of Scurvies. On board the Excellent, much attention was paid to this circumftance. Captain Collingwood tried it in different ways, and kept a diary of the fick lift. Mr. Scott the furgeon, in his report, fays, that " they failed with ten fcorbutics in the lift ; " that this number fluctuated a little while beer " was ferved; but when half beer and half grog, " it increafed to twenty or thirty. When all " grog or wine, it encreafed from forty to fifty- " fix. He concluded, with the uniform tefti- " mony of others, by faying, that the lemons were " a certain cure." The teftimonies of the other

furgeons,

furgeons, viz. Mr. Kent of the Marlborough, Mr. Dods of the Tremendous, Mr. Bell of the Minotaur, Mr. Sibbald of the Nymphe, and Mr. Drew of the Blonde, were to the fame effect *.

⚹ REPORT of the Effects of LEMONS on board His Majefty's Ship Invincible.

"It may be, in fome meafure, neceffary, juft to mention the ftate of health, on board the Invincible, previous to our fail-ing on the late cruize; in order to fhew, in a clear point of view, the excellent effects of the Lemon and Orange, with which we weie fupplied.

The Scurvy (the only complaint to be mentioned here) made its appearance at Spithead in the firft week of April, in a fevere degree. Five patients were fent to the hofpital; others varioufly ill, to the number of eighteen, at the time of failing; four of that number were incapable of duty. By the laft of the month, ten more had applied. Thofe that were worft, took three lemons and one orange, daily; and the others, two lemons. In every inftance, after the third day, and fometimes fooner, they began to recover, and were fhortly well.

In the beginning of May, patients continued to come down, and were treated in the fame manner, with equal good fuccefs; in the courfe of the month fifty-fix had applied, and were re-covered, or nearly fo: in the latter end of the month, the fruit was all expended; but there ftill remained a few gallons of lemon juice, which lafted until the 2d of June. Patients con-tinued to apply, and two of thofe that had been recovering before the lemons were expended, got worfe in the fhort inter-val from the 2d to the 5th, the day on which a frefh fupply of lemons was received. Their complaints were foon checked by the frefh fruit; it was not found that they recovered fo faft

by

From the circumſtances juſt narrated, it will appear ſatisfactory, that my fears were not ground-leſs. The prevention of Scurvy was a great ob-ject in a Fleet, becauſe we are not aware of the exertions of a reſtleſs enemy. But the period was now arrived, when the neceſſity of checking this ravaging diſeaſe became more apparent. It pervaded every ſhip; and as the cauſes which produced it were general in their operation, ſo it was reaſonable to infer, that every ſeaman in the

Fleet

---

by the juice, though it was given, in ſome caſes, to a pint per diem. To this time, thirty-two had applied; and every day two or three, variouſly tainted, had appeared.

"It does not appear, at preſent; that we ſhall have occaſion to ſend any men to the hoſpital for Scurvy, though I am *poſitively* of opinion, that if we had not been ſupplied with the fruit, many of the ſhip's company muſt have ſuffered greatly, or, perhaps, died.—The numbers that complained, and cured, or nearly ſo, are, in April, twenty-eight; May, fifty-ſix; and in June, thirty-two: in all, one hundred and ſixteen.

(Signed)

June 8, 1795.   T. KENNING, Surgeon."

" *Note.* In the ſketch of the good effects of the lemons ſup-plied in the late cruize, I have not mentioned any thing of the advantage our men had in point of regimen. Previous to our ſailing, Captain Pakenham, according to his uſual liberality, purchaſed five guineas worth of onions, and put them under my care. A quantity of freſh mutton was alſo ſent every day to boil with the portable broth. The ſick had wine inſtead of grog. *T. K.*"

Fleet ought to partake of the means of prevention and cure *.

In my reprefentations I took upon me to explain our condition to the utmoft minutiæ of the
caufe, as well as the method moft likely to give
us relief. The Lords Commiffioners of Admiralty, were therefore pleafed to order every thing
which was demanded. The frefh beef was ferved
according to the ufual allowance, and the falt
provifions withheld. A large fupply of lemons
and oranges were ordered to be fent from the Sick
and Hurt Department, equal to our wants. In
addition to all thefe, fallad was ordered to be procured, by way of affifting the other means of prevention and cure. Vegetation was very late, from
the coldnefs of the fpring months; but what
could not be commanded in the gardens round
Portfmouth, was brought from Fareham and
other places. In my official letters, I took care to
inform their Lordfhips, that I had attended the
vegetable ftalls on market-days, as well as the gar-

---

* " You will no doubt obferve with pleafure, the few fcorbutic patients fent to the hofpital, in comparifon with the
number that have complained of that difeafe. This I can attribute to nothing elfe, but the timely fupply of fruit and vegetables, which were obtained by your order, and which had
the moft fpeedy and defirable effect.
                      (Signed)
Robuft, May 31, 1795.   JAMES TURKINGTON, Surgeon."

dens in the neighbourhood, and had there found abundance for our confumption. We were, therefore, from the 31ft of May, fupplied with near five thoufand weight of fallad every day, which was diftributed, under my own eye, to the different fhips at Spithead. The good effects of thefe refrefhments were aftonifhing ; we had only to regret that they were not fent fooner. They do not reft on the official teftimony of the furgeons only ; the whole of our officers attended to the recovery of their people, with every mark of affection : they had fought and conquered together, and were now from habit like perfons of the fame family.

About two hundred men in Scurvy, had been fent on fhore, during the early appearance of the difeafe; but from the time of being fupplied with frefh meat, fruit, and vegetables, it was not 'neceffary to move them from the fhip. The officers and furgeons agreed, that their men were cured in a fhorter time on board. But more urgent reafons required to keep them on board ; defertion was prevented, indolent habits foon learned in our hofpitals, were not fo apt to appear under the eye of their own officers, and as equal juftice would be done to the cure ; our fhips thereby remainded ready for fervice.

Thus, were thofe valuable men fitted and recruited for duty, at an expence to Government

too

too trifling to be mentioned. It was amply repaid, in a very short time, by the capture of three
sail of the enemy's line, in the very mouth of a
French harbour, by Lord Bridport. But it ought
to be remembered, that these ships could not
have sailed a week sooner, from the people that
were laid up with Scurvy bearing a large proportion; to such an extent had it prevailed every
where.

Some deaths happened at the hospital in this
disease; I believe not more than one or two; but not
a single one on board, either at sea or at Spithead.

Although the health of the people was so much
recruited when Lord Bridport sailed; yet it was
plain, that the disposition to Scurvy was not sufficiently corrected, which would have required a
continuance of the sallad for a longer time. But
to guard against the effects of salt diet, and any
accident that might detain the squadron at sea,
every ship was supplied with thirty gallons of lemon juice from the Charon, besides some half
chests of fruit, to those that were most in danger
from a return of the complaint. The London,
Valiant, and Colossus, that had suffered so much,
were now in perfect health.

The weather, for the season, was cold in the
early part of the cruize. On the 10th of July,
the hospital ship parted from the squadron, having received the sick, and forty-five wounded men,

to carry home.   At parting, fome of the furgeons were ferved with an additional fupply of lemon juice from the Charon, left it fhould be wanted before our return.

On the 15th of Auguft we returned to Lord Bridport, having on board a number of fheep, vegetables, &c. I vifited every fhip in the courfe of the day, and found the Scurvy beginning in all ; even the frigates had many cafes of it.

We have feen, in Admiral Waldegrave's fqua-dron, the rapid increafe of Scurvy, as foon as the beer was done ; it was alfo obferved in Lord Bridport's *. I have even remarked at Spithead, fome fhips, that indulged the feamen more than others with fpirits from the fhore, have on that very account a longer lift of fcorbutics, and with more aggravated fymptoms. This fact gives con-fiderable force to my former arguments on the theory of Scurvy.   The ufe of fpirits in the drink

* This is an incontrovertible argument, in favour of my propofal for beer of a ftronger body. A fhip would be able to carry eight weeks allowance to fea : if the cruize or paffage was longer, it might be ferved alternately with wine; by which means, a fhip going to the Eaft or Weft Indies, might prolong it to their arrival at the place of deftination.. We fhould then have no condemned beer ; for a little more hop, and being ftronger in quality, would make it keep. I would have a quart ferved in the forenoon, and another in the evening. In the warm fummer of 1794, the beer that became unfit for ufe in our Fleet, exceeded all precedent for the immenfe quantity.]

of

of the failor, called *grog*, manifeftly tends to abftract the oxygene from the body: from a deficiency of this principle, I apprehend the difeafe is
produced *.

After Lord Bridport's fquadron had been ten
weeks at fea, it then appeared how much the
fafety of our men depended on the citric acid.
There was not a cafe, in which it was given, where
it did not produce a cure in the fpace of a few
days †.

The Robuft, at anchor in Quiberon Bay with
Sir J. B. Warren, was obliged to leave the ftation
from becoming fo over-run with Scurvy. Some
of her people funk under it, from not fending
to the Charon for a frefh fupply of the lemon
juice ‡.

Vinegar was carefully ferved to the meffes of
feamen throughout the fquadron, to be ufed with
the falt meat: yet in thofe fhips where the men

* Obfervations on Scurvy, fecond Edition. Longman,
London.

† " I have no remarks worthy of infertion, except additional proofs of the moft happy effect, from the ufe of lemon
juice: not only in fcorbutic affects of the extremities, but in
pneumatic attacks, when a fcorbutic diathefis prevails.

(Signed)
Pallas, Sept. 20, 1795. R. HARRISON."

‡ See Mr. Turkington's Report in the General Abftract of
Health.

E e took

took it in large quantities, it was not obferved to retard the advancement of the difeafe. Even porter failed to cure. There was now another caufe that accelerated the approach of Scurvy. The fhips had been fo long at fea, that there was a' neceffity for curtailing the allowance of water: the number of fcorbutics increafed in a great proportion; and the London and Prince George were obliged to leave the Fleet.

Thefe fhips made for the firft port, and arrived at Plymouth in the beginning of Auguft. Mr. Smith, of the London, fent eighty cafes of the Scurvy to the hofpital, and thirty in other dif-eafes. Mr. Harris, of the Prince George, fent four men in various complaints. The purfer of the latter fhip was generous enough to fupply fo large a proportion of vegetables, that two hundred fcor-butics were cured on board, and the difeafe ef-fectly fubdued among the crew. The condition of the two fhips, both fecond rates, affords a va-luable leffon on fervice : the London not having the means of curing her people on board, by hav-ing only an acting purfer, who was not empower-ed, lay inactive, for a length of time, in Plymouth Sound; and many of her men never returned from the hofpital. The Prince George, from the bounty of a fingle individual, was thus enabled to cure her feamen in the fhip, and failed for Portfmouth, in lefs than four weeks, in perfect
health,

health, to receive the flag of Rear-Admiral Chriftian, deftined for the Weft Indies. Thus we fee,
as certain meafures predominate, that the Navy
of Great Britain may be occafionally difabled, or
rendered active.——Reafons fimilar to thofe now
affigned, made the Robuft leave fixty men at
Haflar, when fhe failed a fecond time for Quiberon Bay. This is one of the greateft misfortunes
that can befall either a fhip, or the people left
behind *. The fhip's company is replaced by
draughts of raw men from guardfhips, or another
fhip is broke up to recruit the other. The fick
that were left on fhore, when recovered, are fent
to a guardfhip, perhaps not famous for good
order ; and in the end they are fcattered into
ftrange fhips, where they have a new acquaintance to form, and where new officers, perhaps,
inculcate very different modes of difcipline to
what they have been accuftomed : fuch are the
refiftlefs arguments that may be produced, to

---

* The Anfon frigate, Captain Durham, one of Sir J. B. Warren's fquadron, was obliged to go to Plymouth in July 1795,
being overcome with Scurvy. She landed eighty feamen, and
was under the neceffity of leaving a number behind, when fhe
failed for Quiberon Bay a fecond time. Not having received
vegetables while in port, fufficient to fubdue the general taint
among the crew, the difeafe quickly re-appeared ; and had t
not been for very large fupplies of lemon juice from the hofpital fhip, fhe muft either have returned to England again, or
buried half her people in Scurvy.

juftify

juſtify the propriety of ſerving the ſeamen with every comfort in their own ſhips *.

On the 19th of September, Lord Bridport's Fleet arrived at Spithead. The only diſeaſe that was known in the ſhips, was the Scurvy. In the General Abſtract of Health, I have mentioned the number in each ſhip; but every one on board might be conſidered as labouring under ſome degree of the diſeaſe.

Large ſupplies of vegetables were ſerved throughout the whole; and from the quantity of lemon juice being diminiſhed, the ſurgeon of the Charon was directed to purchaſe, at the Iſle of Wight, fifty buſhels of apples, in the immature ſtate. The lemons indeed were now ſo ſcarce, and the conſumption of the juice had been ſo great, that little was left in the kingdom. We were there-

---

* Mr. Perry, then ſurgeon of the Adamant, ſent me ſome Spaniſh *onions*, which hé had brought, in the winter, from Cadiz. In order to preſerve them, he cut the root off to the quick, and applied a ſlight ſolution of lunar cauſtic to the wound, which checked their vegetation ſo much, that oné of them lay in my cabin window for ſix months, without ſhooting. This philoſophical experiment may be eaſily accompliſhed by a hot poker, or any piece of iron, with which the roots ſhould be effectually ſeared. Country people are in the cuſtom of running a red hot knitting needle from the root to the top of the onion, which anſwers better. Theſe proceſſes preſerve this vegetable for a length of time, and as it is one of the moſt uſeful at ſea, they cannot be too generally known.

fore

fore very glad to procure the apples, as a fubftitute. The Royal Sovereign, in particular, reaped much benefit from them, and did not fend a fingle patient on fhore. Mr. Kein found his people well fatisfied : I was indeed glad of the change of fruit ; for I was tired with feeing the poor fellows drinking conftantly of the lemon juice. The cure was every where fo compleat in ten days from the arrival, that we had nothing left, but the flighteft remains, to fhow Earl Spencer, when he vifited the Fleet on the end of the month.

On the 6th of October, the hofpital fhip joined the fquadron under Rear-Admiral Harvey, off Belle-ifle; when the fheep, vegetables, apples, lemon juice, and porter, was diftributed to the fhips for the ufe of the fick. This fquadron put into Quiberon Bay, and lay at anchor there, except a fhort cruize, till the end of December. During this time, frefh fupplies of cattle, fheep, and vegetables, were frequently fent out ; fo that the Scurvy was kept under, or cured when it appeared, with little trouble, by the lemon juice *.

E e 3   Thefe

* " I am opinion, that the lemon juice that was ferved out to our people three or four times a day, diluted with about two ports of water, and a fuitable proportion of fugar; with the timely fupply of vegetables and frefh meat, received while the Fleet remained in the Bay, and iffuing of wine a confiderable part of the time, checked, and in a great meafure prevented,

Thefe fhips came to Spithead on the 2d of January 1796, and with a fquadron, juft arrived, under the command of Vice-Admiral Cornwallis, were fupplied, on frefh-meat days, with a large allowance of vegetables. This allowance was continued to all our fhips, for a fortnight after coming from fea. From this period we may date the extinction of Scurvy in the Fleet.

We have thus feen, from the effects of a cold and rigorous winter, the impolitic and inconfiderate meafure of reducing the allowance of frefh beef, in harbour, and the deftruction of vegetation by froft, a Scurvy induced, fo general in its extent, as to endanger the fafety of the whole Channel Fleet of Britain. The practical inferences to be drawn from this narrative are obvious.

It appears highly expedient, that a Fleet juft come from fea, fhould have the people recruited for future fervice in as fhort a time as poffible.

However

---

ed, our people from being afflicted with Scurvy, and other diforders.   (Signed)

R. Forrest, Surgeon."

Prince of Wales, Jan. 2, 1796.

N. B. Thirty cafes of Scurvy appeared in the Prince of Wales during this abfence of nineteen weeks from England; but there were more in the other fhips of the fquadron.

T. T.

However vague and uncertain the records of naval tranſactions left this point on former oc- caſions; the late occurrences of the Channel Fleet have ſufficiently eſtabliſhed the fact, that Scurvy can always be prevented by freſh vegetables, and cured effectually by the lemon, or the preſerved juice of that fruit.

In the General Abſtract of Health, I have men- tioned, that Dr. Archibald Thompſon, of the Valiant, found, upon comparing his caſes, that not a man who ſhared of the large allowance of lemons and ſallad at Spithead, towards the end of May and the beginning of June, had the leaſt ſymptom of Scurvy during the long cruize with Lord Bridport. Mr. John Smith, of the London, on comparing his caſes, found, that only five of all that he ſent on ſhore to Plymouth Hoſpital, were of the number who partook of the former refreſhments at Spithead, and for theſe he gives reaſons. I think it proper to mention theſe two ſhips, becauſe they ſuffered more than others, from the diſeaſe in April and May. Their re- ſpective ſurgeons took alſo much trouble to aſcer- tain the fact, with ſufficient clearneſs, on my ſug- geſtion*.

Whatever,

* " S I R,
" London, Jan. 6, 1796.

" According to your requeſt, I have examined my books very carefully, and find only five men who had the Scurvy in

April

Whatever, therefore, may be the theory of sea-
scurvy, we contend, that recent vegetable matter,
imparts a *something* to the body, which fortifies it
againft the difeafe : and that in proportion to the
quantity of this *something* imparted, making allowance, at the fame time, for external caufes,
which counteract its effects on the conftitution,
the fymptoms will fooner or later appear. The
prefervative means ought, therefore, to be attended to, and we ought to truft only to the
vegetable acid, when we can do no better.*   The

April and May, fent to Plymouth Hofpital in September,
after more than three month's cruize.   Three of them had a
fcorbutic flux; and two had blotches, &c. on their legs, with
other fymptoms.   One of thefe had but lately recovered from
a venereal complaint, being fix weeks under cure : the other
was always of a fickly habit.

"Two out of the eighty fent to Plymouth Hofpital, in
September, for Scurvy, had been at Hallar Hofpital in April
and May.   One of them was fent for a fcorbutic flux ; the
other had blotches. &c.

I am, S I R,
Your moft obedient fervant,
(Signed)   *John Smith."*

To Dr. TROTTER,
Phyfician of the Fleet."

---

* It is the cuftom of fervice, for the purfer to order his
fteward to buy vegetables, greens, or cabbages, for the frefh
beef broth out of his neceffary money: but there is no fixed
quantity.   There is a neceffity for fixing the proportion, coft
what

juice of lemons long continued, tends to weaken the ftomach and general habit, and produces emaciation in proportion to the length of time it is ufed.

We have alfo found, that the preferved juice is very inferior to the fruit, in its intire and recent ftate; although, I believe, that great attention was paid to its prefervation, and when there could be no fufpicion of vinegar, or other acids, fraudulently added. Three dozen of found lemons did generally as much as a gallon of the juice. Yet we have feen, in Mr. Moffat's report, that he had cured many with the juice that had been near two years fqueezed from the fruit *.

We have, from this practice, eftablished another fact, of the firft importance in naval operations. We have often heard people talk of land air, and land recreations, for the Scurvy. There
is

---

what it may. Why is an article of diet, that is the beft fecurity againft a fatal difeafe, left to fuch uncertainty, or liable to be withheld by any individual, when its price becomes a little higher than common ?

* This general Scurvy in the Fleet, would have afforded a fine opportunity for trying the effect of oxygene air in the cure. But, as I was the firft who ventured to publifh the Theory, there was a danger of being accufed of a predilection for fpeculative opinions, had I applied to Government for an apparatus.

is not at prefent an officer in the Fleet, that, in doing juftice to either his people or his country, would prefer the cure out of a fhip. Nay, there is often the moft urgent neceffity for keeping them on board till they acquire a certain degree of ftrength. In the very weak ftage, a fcorbutic patient cannot bear the external air, which has been long obferved, and recently confirmed, by the five men dying in the boat belonging to the Prince of Wales, between the Downs and Deal Hofpital.

Thofe who may confult thefe pages, and are acquainted with the opinions, which I formerly publifhed, on the Theory and Practice of Sea Scurvy, will fee, that this immenfe field for obfervation, which the Channel Fleet has afforded, feems to ftrengthen all my old conclufions. Nothing remains for me to retract. The whole has been confirmed by the experience of a great number of furgeons, of approved abilities, in the profeffion. Dr. Beddoes, in his remarks on my work, laid much ftrefs on the impure air of fhips in producing Scurvy. If this had fo much effect, furely it would have counteracted the cure, when the feamen remained on board : but that has not been obferved. The furgeons generally remarked, a very great difference on the fecond and third day; and a week was long enough for to compleat

pleat the cure. The difcipline of the officers in the Fleet, in whatever related to health, has rendered them famous. To have thought of foul air, as a caufe of the Scurvy, when it appeared in the Royal George and Queen, would have been the laft refource of a phyfician, inveftigating caufes, who had witneffed the admirable fyftem of duty practifed by Captains Domet and Bedford.

I fhall conclude my obfervations on this difeafe, by introducing a letter from Mr. Baird, furgeon of the Hector. This fhip left the Channel Fleet in May, when our fufferings were great.

" Hector, Spithead, Dec. 4th, 1795.

" S I R,

" As I confider the Navy indebted to your exertions, for the very valuable inftitutions of lemon juice ; I fhould think I failed in my duty, if I did not communicate to you, the wonderful benefit derived from it in the Hector.

" On my joining this fhip, in May laft, I found her under orders for foreign fervice ; our deftination fuppofed to be the Eaft Indies. Several of our fhip's company were labouring under Scurvy, in an advanced ftage, all of whom I fent to the hofpital, previous to our failing. But, as I had

had great reaſon to ſuppoſe, that the ſcorbutic
taint was general, I rather felt diſcouraged at the
idea of encountering ſo long a voyage, under ſuch
circumſtances. Indeed my fears were ſoon confirmed ; for we were but a few days at ſea, when
the ſick liſt was conſiderably increaſed with ſcorbutic patients : ſome with their gums highly
putrid, legs and thighs much ſwelled, hams contracted, and ſo very ill, as to render them totally
unfit for any kind of duty. The ſmall beer not
being expended, the inſtructions from the Sick
and Wounded Office did not exactly warrant my
iſſuing lemon juice and ſugar. But, as it ſeemed
the only probable remedy, I ſolicited Captain
Montague to loſe no time in giving it to the
ſhip's company, in the quantity directed, as a
preventive ; to be mixed with a proportion of
water, as ſherbit ; and alſo to allow me to iſſue
it, in any quantity I might think proper, in bad
caſes.

" I began with giving the lemon juice, in the
quantity of an ounce and half daily ; and, encouraged by the material change I perceived in
about four days, I increaſed it to three or four
ounces *per diem*; always taking care to join a
ſufficient quantity of ſugar, to prevent it from
irritating the bowels : in twelve or fourteen days,
the worſt of them were able to return to duty ;
every

every fymptom being then removed, except fome
flight degree of ftiffnefs in the hams, which gradually wore off.

" That the lemon juice was equally ferviceable
as a preventive, as I think evident from ,the following obfervation; for the firft month, new patients frequently complained, but after that time,
the only fcorbutics we had, were men preffed
out of merchant fhips returning from long foreign
voyages.

" We failed with the outward bound Eaft
India trade on the 24th of May laft; and returned the 19th of November, with the homeward bound trade; having been as far as 28°️ and
a half fouth; about three weeks in harbour from
the time of failing till our return; and part of
that time at St. Helena, a place well known for
its barrennefs, and the very little refrefhment
which it produces: We have returned to Spithead without lofing a man by Scurvy; nor have
we had occafion to fend a fingle man, in that
difeafe, to the hofpital.

" When I confider the alarming progrefs
which the Scurvy was making among the Hector's
fhip's company, previous to the adminiftration of
lemon juice as a preventive; the fudden check
that difeafe met with afterwards; the powerful
effect of the acid in very bad cafes; I think I

3 will

will not be accufed of prefumption, when I pronounce it, if properly adminiſtered, a moſt *infallible remedy*, both in the cure and prevention of
Scurvy.

I am, SIR,

Your moſt obedient,

and humble ſervant,

(Signed)   *A. Baird.*"

To Dr. Trotter."

THE MEANS used to eradicate

A MALIGNANT FEVER,

Which raged on board His Majesty's Ship BRUNSWICK,
at SPITHEAD, in the Spring of the Year 1791:

With some short OBSERVATIONS on the most probable Means
of preserving the Health of a Ship's Company.

By SIR ROGER CURTIS*,
Then Captain of the Brunswick.

THE health of the Crews of His Majesty's ships, is a consideration of so important a nature, that whatever may contribute to its preservation, or to the removal of contagion, when it unhappily exists among them, are circumstances which merit the serious attention of every officer. Those who are conversant in our maritime history, must have regarded, with horror, the dreadful havock which disease, for a long continued series of years, made amongst the seamen of this country.—They will have observed, that disease carried off an hundred times more than fell by the hands of the enemy. During those unhappy

---

* Captain of the Fleet to Earl HOWE:

" To Souls like these, in mutual friendship join'd,
" Heaven dares intrust the cause of humankind."
Addison, on Marlborough and Prince Eugene.

periods,

periods, it is true, that medical ſkill had not reached the degree of perfection it now poſſeſſes ; but the chief cauſe of the dreadful calamities which befel our Fleets, was the want of that order, cleanlineſs, and internal œconomy, which is now more generally obſerved. It is now proved, that a due attention to theſe circumſtances operates moſt powerfully, in all ſituations, towards the preſervation of health. But, notwithſtanding the ſtricteſt adherence to the wiſeſt precautions, diſeaſe but too often finds its way amongſt us. Fevers, of the moſt infectious and dangerous kind, frequently rage in our ſhips. They are ſometimes generated there by a want of cleanlineſs in the ſhips, and in the people; but they are more generally propagated by the introduction of infected perſons.

When contagion has once taken root in a ſhip, the different parts of it, as well as the perſons and the cloaths of the crew, become highly infected, and it cannot be removed without great labour and perſeverance. Under circumſtances, when neither the ſtores nor the crew can be removed from the ſhip, the difficulty of oradicating infection is greatly augmented ; but even thus ſituated, it is *poſſible* to be effected. It requires, however, great and unremitting pains. The ſlight and ordinary modes of fumigation, by correcting the air, are ſerviceable in the *prevention* of ſickneſs ; but when
contagion

contagion is eftablifhed, a more powerful application of it, and other means muft be adopted.

The Brunfwick was afflicted with a putrid and
highly infectious fever, when lying at Spithead in
the fpring of 1791, which raged fo violently, that
frequently ten or fifteen men would fall down in
it in a day, and more than one hundred and fifty
were in the hofpital at Haflar at one time. Its
progrefs in the fhip was however at laft arrefted,
and the means made ufe of are hereafter related,
that they may be followed by others under fimilar
circumftances, if they are deemed to be deferving
of notice.

As feamen have great reluctance in complaining
when they find themfelves but flightly indif-
pofed, and it being very material that infected
perfons fhould, as fpeedily as poffible, be removed
from the body of the fhip's company, to impede
further communication of the difeafe, as well as to
facilitate the cure of thofe attacked, by an early
application of medicine, great attention was obferved by all the officers, in immediately reporting
every man who appeared to have the fmalleft indifpofition, whether it was difcovered by day or by
night. The whole fpace under the forecaftle on
the larboard fide, including the round-houfe, was
appropriated to the fick, and the obtrufion of any
other perfon abfolutely prevented. To this place
every perfon was removed the moment it was dif-

covered

covered that the difeafe had feized him, and the primary remedies towards cure immediately applied, from whence, as fpeedily as could be, he was carried to the hofpital, care being had that every thing belonging to him was fent with him.

Moifture operating more powerfully than any other caufe in the production of difeafe, as well as in the propagation of it, our firft care was the endeavour to remove all humidity, and foulnefs of air.

The well was baled out, fcraped and fwabbed till entirely dry, and then a large fire was kept burning in it for feveral hours every day, fo that the fmalleft dampnefs therein was not fuffered to remain.

The hold had the upper tier of cafks removed from it, and fent on fhore. Three fires were then kindled in it, and kept burning for many hours every day, confining the fmoke as much as poffible, and occafionally fhifting the fires from place to place; and in the fuel made ufe of, as many empty tar barrels were confumed as could be collected for the purpofe; at other times wood, and occafionally coals, intermixed with fhakings of tarred ropes. Every precaution being conftantly taken to prevent accident. When the fires were extinguifhed, the gratings of the hold were removed, and the windfails let down.

The

The horlop was conftantly kept as clear as poffible of every thing that prevented a free circulation of air, and a fire placed fometimes in one part of it, and fometimes in another.

The cockpit, fteward room, and bread room were treated in the fame manner.

The doors of all the ftore rooms were occafionally thrown open, and the ventilators worked unremittingly day and night.

Three fires on each fide the 'tween decks were kept burning almoft the whole day, and thefe were from time to time fhifted to every part of it. The manger was cleared of all manner of lumber, and a fire occafionally placed therein. The deck was feldom wafhed, and never but when the weather was fuch, that the people could remain upon the upper deck, until it was perfectly dried by the fires, and the natural current of the air; nor was any perfon whatever permitted to go below, under any manner of pretence, until the general permiffion for it was given. When the deck was not wafhed, it was kept perfectly clean by other means; and flops about the decks, and every fort of dampnefs was fpecially guarded againft. The fides, beams, carlings, the deck overhead, and every part of the 'tween decks were white-wafhed, twice or thrice, during the courfe of the diforder.

F f 2

Fumigation

Fumigation in the hold was thus conducted :—
four half tubs with ſtands in them were diſpoſed
therein. In each of the tubs was placed an iron
pot, into which was put about two pounds of
brimſtone tied up in a piece of canvas. The gratings were laid, and ſo cloſely covered with tarpaulins, old hammocks, ſwabs, &c. that none of
the ſmoke might eſcape. When every thing was
prepared, a red hot loggerhead, or iron fid, was put
into each of the iron pots, which ſet fire to the
brimſtone; and the men performing this ſervice,
immediately leaving the hold by a grating of the
main hatchway being kept open for the purpoſe,
the hatches were entirely cloſed.

It was the cuſtom, to fumigate the hold, horlop,
and 'tween decks at the ſame time; but as we could
not be furniſhed with a ſufficient quantity of
brimſtone, to make uſe of it in all the different
parts of the ſhip at the ſame period; it was uſual,
therefore, to uſe the *brimſtone* in the hold, horlop
and 'tween decks in rotation; and where the
brimſtone was not applied, there were ſubſtituted
what are called devils, made of powder wetted
with vinegar. In thoſe parts of the 'tween decks
leaſt acceſſible to air, and where conſequently
there is a greater degree of contagion, the flaſhing
of powder from piſtols is attended with very good
effect; for the ſhock of the exploſion aſſiſts very
powerfully

powerfully in difperfing the infectious matter attached to the timber of the fhip.

During the fumigation, the men's hammocks
were all hung up in their places, with their mat-
traffes and blankets fpread over them, and all their
fpare apparel was fo difpofed of upon the guns,
&c. as to receive the full effect of the fumigation;
and the cloaths, which the men wore upon deck
during the time of one fumigation, were changed
upon the next, and placed below, that all their
things might receive equal purification.

The gratings on the main deck were laid and
covered, and with fuch care, that no fmoke could
efcape, and the pots were carefully barred in. The
brimftone in tubs, or the devils, with other fafe
precautions, were difperfed about the decks, and
then lighted; the perfons who did it efcaping
upon deck, and clofing the hatchway after them,
the operation was completed.

The fmalleft crevices of the fhip were pervaded
by the fmoke and effluvia of the brimftone, and
affected every part of her in a powerful and
aftonifhing manner.

Three hours were generally fuffered to elapfe
before the gratings were uncovered, and the ports
opened; and a free circulation of the air for a
very confiderable time was afterwards neceffary,
before a perfon could remain below without inconvenience. The whole of the hull of the fhip,

and

and every thing therein, animate and inanimate, was ſtrongly impregnated with the fumes of brimſtone, and to ſuch a degree, that it was perceptible when to leeward of the ſhip, at a conſiderable diſtance from her.

In damp weather theſe fumigations were practiſed every day, and never leſs than three times a week. The fires were continued daily.

The ſick birth was attended to with the ſame ſolicitude, to impede and eradicate infection, as has been deſcribed in reſpect to the other parts of the ſhip.

Nor were the perſons and apparel of the men diſregarded. Every man in the ſhip was waſhed from head to foot with warm water and ſoap, and more than even our uſual pains were taken that they ſhould be cleanly in all reſpects. If any old and uſeleſs cloaths were found, they were thrown overboard. Such ſerviceable apparel as was diſcovered the leaſt filthy, was waſhed and fumigated, and the men were forbid to wear woollen trowſers. On fine days, the whole of their bedding was hung upon lines between all the maſts, and on the rigging, and expoſed thus to a free ventilation for many hours, and their cloaths of every kind were treated in the ſame manner.

The recovered men, returned from the hoſpital, were treated upon their coming back to the ſhip with great precaution. Having received at the hoſpital hofpital notice of their recovery, and the intention of their return, a careful petty officer was fent thither to fee all their cloaths and bedding well aired, by being fpread abroad for two days, and well beat and cleaned, previous to their coming to the fhip. Upon their arrival on board, every man was wafhed in warm water, and with foap, and an entire change of cloaths was then put upon him; all the reft of his apparel and his bedding were immediately fumigated with brimftone, which was performed by fufpending it over the fumes iffuing from an iron pot, placed in a half tub, in a convenient place under the forecaftle.

Such were the means made ufe of with punctual and unceafing perfeverance. They certainly were attended with no little labour, but the fever with which we were afflicted having been entirely fubdued, the gratification, arifing from the reflection that our endeavours were crowned with fuccefs, was the moft ample recompence for all our trouble.

THOUGHTS

ON THE MOST PROBABLE MEANS OF

Preserving the HEALTH of SEAMEN.

———

A CHAIN of calamitous Circumftances but
too frequently occurs, to endanger the
Health of our Seamen; and on fuch occafions the
utmoft endeavours of the Officer to prevent it
altogether, proves frequently ineffectual. But except in thofe extraordinary cafes, prevention is
more in the power of officers than feems generally
to be imagined. It would fave the lives of many
men, and be otherwife productive of very beneficial effects, if fires were, in the day time, to
be placed in the hold and other parts of fhips,
for fome weeks previous to their being put into
commiffion. And the difficulty and expence attending the meafure are really contemptible, to
the good that would refult from it.

Great care fhould be taken, that the ballaft received on board fhould be clean, frefh, and as

good

good in its kind as can be procured. The Thames ballaft is very unfavourable to health; for befides being naturally foul, it is full of frefh water animalcula, which being deftroyed in the fhip, ftink, and become very offenfive.

It is fometimes the practice to line the gratings of the horlop with thin deal, to prevent any filth from falling into the hold : but it is a method highly pernicious; for the preclufion of air, which is thereby occafioned, fills the hold with putrefcency, induces dreadful difeafe amongft the people, and deftroys the fhip. Every method therefore fhould be taken to caufe a circulation of frefh air in the hold, by the ufe of windfails, and occafionally by fire.

The well fhould be kept as dry as poffible; and if the tightnefs of the fhip will allow it, it fhould be cleanly fwabbed out every day. If it cannot be kept properly dry, water fhould be let into it every evening, and pumped out the following morning, and in the day-time fires frequently made in it. Fatal accidents frequently happen from the bad ftate of the well, in cafes where men are obliged to go into it upon duty; but where even thefe do not occur, very great is the injury done to the health of a fhip's company, by want of due attention to the ftate of the well. It is a very proper precaution to try a lighted candle in the well before any perfon be permitted

mitted to go into it, if there is the leaft fufpicion
of the air therein being foul.

Upon the firft outfit of a fhip, the men received
on board her, fhould be examined with the moft
fcrupulous attention, that difordered and infectious, or foul ulcerous perfons may not be admitted. Such of their cloaths as are foul, and of
little worth, fhould be deftroyed, and the refidue
wafhed and fumigated; and their perfons fhould
alfo be thoroughly cleanfed, by caufing them to be
wafhed with foap and water from head to foot.
Their hair, if neglected and filthy, fhould be cut
fhort, and if neceffary their heads fhould be
fhaved. Without thefe precautions, the feeds of
future malady may be diffeminated in the fhip,
and diforder may afterwards make dreadful
havock amongft the crew. And at no time
fhould any ftrange men be received on board,
without fimilar attention.

Order and cleanlinefs are the officers' moft
powerful refources for the prefervation of health;
but there are a thoufand leffer circumftances to
be attended to, which materially contribute to fo
defirable an object. The nature of cleanlinefs too
is often mifunderftood; and I know of nothing of
that kind which is fo much miftaken, as the too
frequent and indifcreet drenching the decks, and
more efpecially thofe where the people fleep, with
water, and particularly in cold latitudes during
the

the winter. By this means I have known dreadful
ſickneſs *introduced*,—and I have known it *removed*
by a contrary practice. It would be deemed ex-
travagant to advance an opinion, that the decks
ſhould *never* be waſhed; but I feel no reluctance
in making a direct aſſertion, that it were far better
that they ſhould not be *waſhed at all*, than with
that want of diſcretion and precaution, which ſo
generally prevails. It is an error that has cauſed
the deaths of thouſands! Certain it is, that the
decks cannot be kept too clean; but they ſhould
be made ſo by other means than waſhing, except
the weather be ſuch as will ſoon cauſe them to
dry, or that you have the means of drying them
by fires. This obſervation applies to every deck
in a ſhip; but in a particular manner, to thoſe
where the people chiefly reſide, or from whence
the humidity particularly affects them: nor ſhould
they ever be permitted to go below after a waſhing, until the decks are perfectly dry; for it is a
fact univerſally admitted, that moiſture is the
chief prediſpoſing cauſe to almoſt every malady
with which a ſeaman is afflicted. It particularly
induces ſcurvy and putrid fevers. Seamen are
naturally indolent and filthy, and are merely in-
fants as to diſcretion, in every thing that regards
their health. They will aſſiſt in waſhing decks,
and ſit the whole day afterwards, though wet
thereby, half way up the legs, without ſhifting
                                            themſelves,

themfelves, to the great injury of their health.
They fhould therefore be compelled to put off
their fhoes and ftockings, and roll up their trowfers
on thofe occafions, which will not only caufe their
feet to be dry and comfortable the reft of the day,
but neceffarily caufe a degree of cleanlinefs which
otherwife would be difregarded. The practice
which has lately been adopted of having ftoves
with fires placed occafionally in thofe parts of the
fhip where the men refide, and in others fubject to
humidity, is of the utmoft importance to the
health of the people, and fhould never be omitted
in damp weather.

Great pains fhould be conftantly taken, that
the men are cleanly in their perfons, and that they
are furnifhed with all neceffary cloathing. And
the method now fo generally adopted of appro-
priating the fhip's company to the care and fuperintendance of the feveral officers, renders the ac-
complifhment of thefe neceffary and important
requifites extremely eafy.—Seamen have a cuftom
of dreffing themfelves to undergo infpection at
ftated periods, while at other times they are covered with rags and naftinefs. They fhould be
compelled to keep their trowfers and other cloaths
clean, how much foever they may be worn; and
that they may have no excufe for raggednefs, there
fhould be a taylor to every divifion, whofe fole
employment, under the direction of the officer,

fhould

should be in keeping in order the cloaths of the men belonging to it. Whenever any payment is made on board, the officers of divisions should take care that their men do lay in a sufficient stock of cloaths, with soap, and such other articles as may be necessary for them, before they are allowed to squander any of their money in dissipation. And where these payments are made upon the eve of the ship's going to sea, the men should be induced to lay in a stock of onions, potatoes, and even tea and sugar. It is inconceivable, to those who have not seen the good effects of it, how much such attentions preserve the health of a ship's company.

The bedding of seamen is in general too little attended to; for as they sleep without sheeting, their blankets must unavoidably become offensive, and injurious to their health. Their bedding should therefore be frequently well aired, by being hung upon lines fixed for the purpose; and their blankets should be washed as occasion may require.

Their hammocks should be kept clean; but whenever they are washed, they never should be permitted to sleep in them until they are perfectly dry; for they had better spread their bedding on the deck, than lie in a damp hammock. Whenever it can be done, there should be a change of hammocks in use for the ship's company;

pany; and this may be managed without the leaft additional expence to government. But what are expences, compared to the health and comfort of the people! The hammocks fhould be got up whenever it is practicable; to admit of a free circulation of air below, and this may be generally done, owing to the excellent hammock coverings which are now allowed by government. But when the weather is thoroughly dry, the hammocks fhould be frequently uncovered in the nettings, in order that they may be benefitted by the free accefs of the air. The cleanlinefs of the people's bodies fhould be particularly attended to, as well as their cloaths and bedding. They fhould therefore be compelled to wafh themfelves in tubs allotted for that purpofe, which will not only contribute to prevent illnefs, but will alfo act as a bracer, and render them lefs liable to catch cold.

Having but little judgment of what is fitting for them in any fituation, they fhould not be permitted to go too thinly clad in fevere weather, nor too warmly when it is hot. They are too indolent to fuit their drefs to circumftances, unlefs they are forced to do it, nor is any thing more common than to fee fome of them with a pair of thin linen trowfers on in the feverity of winter, and a pair of greazy woollen ones in the hotteft. weather.

When

When their watches expire in rainy weather, they fhould be obliged to take 'off their wet fhirts before they get into their hammocks, which, from lazinefs as well as fatigue, they will not do, but by compulfion. Nothing can be more pernicious than going to fleep wrapped up in wet linen, and it caufes alfo their bedding to be damp and unwholefome for fome time afterwards.

They fhould never be allowed to fleep upon deck during their watches, a cuftom too prevalent, but which is always greatly injurious to the health of the people, and more particularly fo in warm climates, where the dews are profufe, than even in colder latitudes.

Many officers are of opinion, that it is advifable to put the men to three watches in preference of two, when it can be done, from a fuppofition that the greater degree of reft thereby given them, it muft be conducive to their health; and this would certainly be the cafe, were they to fleep in a pure air. But as the places where they reft are particularly clofe and confined, during the night efpecially, and the air rendered foul by a number of people crouded together, there are thofe who think it better for them to have fhorter portions of reft, by being at two watches, than to continue for eight hours together fweltering amidft an highly corrupted air. And great care fhould

fhould always be had, that the people fhould be fo
birthed, that the portion of the crew off of duty
fhould be difpofed generally over the fhip, and
not arranged together in a particular part.

Nothing is more commendable, or has a better
effect on the people, than parade, order, and re-
gularity in a fhip of war; but thefe things fhould
always be fubfervient to the health of the men.
In line of battle fhips, the ports being up and the
guns run out, has a fine appearance : but, in this
cafe, the health of the crew fhould be confulted ;
for in cold damp weather, and particularly when
the fhip is broadfide to the wind, the expofing of
the men to a current of air on fuch occafions, and
when too they are in an inactive ftate, is extremely
detrimental to their health, and brings on a train
of difmal diforders. At thefe times the weather
ports fhould be fhut. A frame to fit every port,
with double bunting ftiched to it, is an excellent
method to correct the evil confequences of too
great a current of air paffing through the fhip at
improper feafons. It prevents the ill effects of
too much air, but admits enough for fufficient
ventilation.

It has been obferved, that the men belonging
to the boats are more frequently difeafed than any
other equal proportion of the fhip's company.
The reafons are obvious : they are more expofed
to the weather. They frequently get wet on
going

going on fhore, or become fo there, and in that condition they often remain many hours, waiting upon the beach for fome giddy and unthinking officer, who, amidft his own enjoyments, thinks too little on the fufferings of the boat's crew. Upon the return to the fhip, inftead of care being taken that they fhift themfelves, and are made comfortable, they pafs unnoticed, and are fuffered to continue in that condition until their cloaths dry upon their bodies.

The boats crews of a fhip are generally ragged as well as fickly. They too often fell their cloaths to buy *liquor*, but too often alfo they part with them to buy food, on account of their being un-neceffarily kept on fhore, to the lofs of their re-gular meals with the reft of the fhip's company. It may with truth be faid, that the inattention to the boat's crews of his Majefty's fhips, is amongft the principal irregularities in the navy that require correction; for it deftroys their health, and is a great caufe of defertion.

Amongft the various omiffions which contribute to the injury of the health of the crews of his Majefty's fhips, nothing feems more extraordinary than the general neglect there is of working the fixed ventilators. Thefe fhould never ftand ftill for a moment. A boy is capable of working them, and it may alfo be made a little extra duty for fuch as may be guilty of very flight offences.

To judge of the utility of them, one has no more to do, than to vifit any of the enclofed parts of the fhip, with which they have connection, where they have not lately been ufed; fet the ventilators a going for a few hours, and vifit the place again, and the difference of the ftate of the air will be found as obvious to the fenfes, as the diftinction between extreme heat and extreme cold. The ventilator in the foremoft part of the fhip fhould be particularly attended. The air in the ftore-rooms there fituated, efpecially the gunner's, becomes fo foul, as frequently to be fcarcely refpirable, and every thing in them decaying or rotten. The powder too is greatly injured by the humidity and foulnefs of the air, and caufes the manufacturer to be blamed for its bad quality, where the diminution of its ftrength has arifen from the want of due attention to it in the magazine, where feldom any pains are taken to correct the air, and where fometimes the powder remains during a whole year, without the barrels containing it being changed in their pofition. The feveral materials of which powder is compofed have, and more efpecially in damp fituations, a natural tendency to feparate from each other. Where, therefore, the barrels remain a long time in one pofition, the faltpetre defcends towards the bottom, the due proportion of parts is deftroyed, and the powder is good for nothing.

9   I am,

I am, therefore, perfuaded that the clamour againft the powder, with which his Majefty's fhips have been furnifhed, has not always been well founded, and that the bad quality perceived in it has generally arifen from a want of due care in its prefervation. I have feen a very great number of trials made of the ftrength of various forts of powder, made in different nations, and the refult of thefe trials was, an unequivocal decifion, that the Britifh powder was fully as good, if not in general the beft, made in Europe. It is clear, therefore, that the barrels containing powder fhould be turned at leaft every two or three months, and particularly when in fo clofe and damp a place as the magazine of a fhip of war.

The fore mafts of fhips are found more generally decayed than either of the others, and this is, undoubtedly, occafioned by the foul and humid air, with which the commonly unfound part is almoft always furrounded. Motives of fafety will, perhaps, continue to prevent thefe evils from being perfectly removed; but they may certainly be much remedied. I have often thought, that the outward doors of the fore ftorerooms, might have a fcuttle in them, covered with copper, punched full of holes, which would admit a confiderable portion of frefh air, without the leaft danger of accident. And, as the ftate of the weather, and other circumftances may ad-

Gg 2

mit,

mit, the doors of the ftore-rooms fhould be thrown open for a confiderable length of time during the day; and to prevent accident, an officer fhould conftantly ftand before them until they are again fhut. Such ftores, as may with fafety be removed, fhould be occafionally got upon deck to be aired, which will greatly tend to their prefervation, as well as give the more free admiffion of frefh air into the ftore-rooms. Thefe remarks are, indeed, fomething digreffive from the profeffed plan of this little work; but it is hoped they will not be thought fo much fo, as to be deemed entirely mifplaced.

The windfails, at every hatchway, fhould be conftantly ufed, when the ftate of the weather will admit of it, and they fhould reach below the horlop deck; for as the hold is unavoidably fubject to confined air, in a confiderable degree, every poffible means fhould be taken to introduce frefh air into it; and it is too little confidered, how many caufes of difeafe originate in a want of due attention to the hold.

As a preventive againft ficknefs, as well as when contagion rages, the white-wafhing the interior parts of the fhip, twice or thrice a year, would, by removing noxious effluvia, greatly contribute to the health of the people.

It is much to the credit of government, that the feveral articles of diet, provided for the navy,

are

are generally of the beft qualities : but notwithftanding this generous care of its fervants, it cannot be denied, that owing to the nature of fome
of the articles allowed, and the unfuitable proportions of others, there is abundant room for
amendment in the mode of victualling the navy;
and it fhould feem, that improvements therein
may eafily be made, which, without additional
expence to the ftate, would greatly conduce to
the health of the feamen.   Deviations from the
ufual articles of victualling were adopted, in various inftances, during the laft war ; and the good
effects of it are too well known and remembered,
to be here particularized.

The articles of cloathing fupplied, deferve alfo
new regulations ; for the fame proportion of the
feveral articles are iffued to the fhips, whether
deftined to a frozen, or to a burning climate.

The want of foap, abroad particularly, is very
generally the caufe of great filthinefs in the perfons and the apparel of feamen.   It would greatly
tend to remove thefe evils, and confequently
contribute to the prefervation of their health, if
this article were to be fupplied by government,
and iffued by the purfer to the people, in fuch
quantity per month as may be deemed neceffary.

At fome fit period, it is probable, thefe important fubjects will meet with the confideration of
government.

Touching the articles now supplied to ships, which relate to provisions, there appears to be nothing so injudiciously used as vinegar. This is a very useful article wisely distributed; but the general practice, at present, is to issue it to the ship's company, in considerable portions to every mess, once in two or three months; but the men having no proper means of keeping it, it is lost by accident, or properly consumed in the course of a day or two, and they have no more until the next serving. I have seen the following mode of supplying vinegar made use of with the happiest effects: a small cask of it has been slung up under the half deck, and, to prevent waste, in care of a centinel. The men had access to it whenever they pleased, and, though they had thus a constant use of it, there was never want of vinegar in the ship.

Whenever ships touch at places where lemons or limes are to be obtained, a considerable quantity should be procured for the use of the ship's company; and, to prevent the loss that may be sustained by their rotting, it is advisable to express their juices. The juice should remain for twenty-four hours in a tub; the scum being then carefully removed, it may be put into a clean cask, and with the proportion of two or three gallons of spirits to a hogshead of juice, it will be preserved for a very considerable length of time.

Thus

Thus furnifhed, a fhip may bid defiance to the fcurvy; and the application of it will be found not a little ufeful as a remedy for other afflictions.

Refpecting water, fome very valuable obferva-tions are to be found in Dr. Blane's publication relative to the Difeafes of Seamen, to which the reader is referred. But it may not be amifs to fay, that no fhip fhould go to fea without one of OSBRIDGE's machines for fweetening it, which is found, by experience, to be the moft fimple and efficacious mode of purifying water that has hi-therto been invented : nor fhould a drop of water be ufed by the men, while a fhip is at fea, without having been previoufly purified by it.

Ships are frequently obliged to have communi-cation with the fhore, where the country is cover-ed with woods and marfhes, and where it muft be, unavoidably, unhealthy. In thefe fituations, no man fhould, on any account, be fuffered to pafs the night on fhore; nor fhould they leave the fhip for the purpofe of wooding or watering too early in the morning, nor ever before they have breakfafted. And it would greatly preferve their health, if before they went upon thefe duties, they took each a fmall quantity of fpirit in which bark had been infufed. They fhould alfo always be brought off to the fhip before fun-fet. Where the crews of fhips *muft* of neceffity be employed in wooding and watering, the rules and precau-

tions

tions above recited will be found highly beneficial; but in unhealthy climates, thefe duties fhould be performed by the natives hired for the purpofe, and particularly on the firft arrival of the fhips.

I beg leave to relate a fact, in confirmation of the utility of the above precautions.   In a voyage down the coaft of Guinea, in the Affiftance, in the year 1762, we had fcarcely a man indifpofed. We wooded and watered at the Ifland of St. Thomas; and, with a view to expedition, a tent was erected on fhore, in which the people em-ployed on thefe fervices were lodged during the night.   On the Middle Paffage, every man that flept on fhore died; and the reft of the fhip's company remained remarkably healthy !

To the humane and learned Dr. Lind we were firft indebted for a moft valuable work, written profeffedly with a view to preferve the health of our feamen; and Dr. Blane, the prefent Phyfician of the Fleet, has very fuccefsfully directed his eminent talents to the fame important end.

As an individual fincerely attached to my pro-feffion, and ardently folicitous to adopt every rule of conduct, by which the more laborious part of it may be rendered healthy and comfortable, I beg leave thus to offer them the tribute of my unfeigned gratitude.   The fame fentiments of acknowledgment are due to them from every well-wifher to this country.   They have not, in-

deed,

deed, directed fleets on the day of conqueft, but they have eminently contributed to preferve the health of thofe by whom victories are obtained. And I fcruple not to fay, that their country has been more benefited by their labours, than by the greateft victory that was ever atchieved! It has been wifely faid, that the fatherly care of a commander is the *Seaman's beft Phyfician.* To know, therefore, how moft ufefully to direct his benevolent and humane difpofition, the works of thefe refpectable men fhould occupy the firft place in the library of every Captain of a man of war; for a careful perufal of them, will teach him the way to render healthy, comfortable, and happy, the men over whom he is appointed to command; who are entitled to kindnefs in return for obedience, and who look up to him as their guardian and protector.

A P P E N D I X.

The TREATMENT of RECENT VENE-REAL INFECTION.

ADDRESSED TO OFFICERS.

IT is not the intention of the prefent addrefs, to make every man his own phyfician : that, the author well knows, is equally hazardous and impracticable. This is a matter, however, from its peculiar nature, on which young men fome-times read books ; and Officers of our Navy do it as well as others. But the reading of books on the Venereal Difeafe, as I have frequently feen, has not been from curiofity alone; there has been a defire to think and act for themfelves ; and we often hear people who have not made phyfic their ftudy, giving their directions on the method of prevention and cure, in a tone of confidence and authority, that befpoke a long acquaintance with the fubject. To perfons of this defcription, a little feafonable advice may be ufeful. I am

the

the more inclined to do this, as many officers of the fervice, have been accuftomed to treat themfelves, in the early ftage of the complaint, according to my directions; by which means they believed that their conftitutions were faved from the ravages of a dangerous difeafe, by a very fhort confinement and pain.—If there was any more urgent motive for my treating the fubject in this manner, it would be gratitude for a thoufand tokens of friendfhip, that no diftance of time or place will ever obliterate from my memory.

It has fallen to my lot, to fee the Venereal Difeafe in as great a variety of cafes, and among perfons of very different difcriptions, as has ufually come to the fhare of moft phyficians and furgeons. I alfo think, as a navy furgeon, I have had fome advantages over gentlemen who practife on fhore, and who have not their patients fo much under their eye, as we have in fhips.

It would be well if a fafe method of treating the Recent Venereal Infection, could even be extended to the feamen : for although the abolition of the fine, for the cure, has done much in making them difcover their complaints early, yet we have known a degree of modefty in fome of them, independent of other confiderations, prevent them from applying to the furgeon, left their names fhould be handed in the fick lift to the Captain. We are now told that the fale of mercurial preparations

parations in the apothecaries fhops in Portfmouth,
and elfewhere, has diminifhed in an uncommon
degree, fince government remuncrated the furgeon for the cure ; and may we not fuppofe that
many valuable lives will be faved in confequence.
I have frequently known feamen enter on board
a man of war, for no other reafon but the cure of
the Venereal Difeafe, which they were not able to
pay for, on fhore ; and if that was the cafe formerly, how much more encouragement does it
hold out now, when they are cured for nothing.
We may therefore confider it, not only as one of
the moft humane and juft alterations which could
be devifed, but as one of the moft popular meafures which the Admiralty could adopt. The
herd of quacks and itinerant practitioners who
frequented the fea-ports, and preyed on the credulity of our men, have alfo taken their departure from the failure of bufinefs. It was a grievous reflection to think that a failor often paid fo
high as five guineas for medicines, while the difeafe, in the mean time, was gaining ground, and
for which he was obliged at laft to go to an hofpital. But there is another advantage, no lefs
humane, derived from the change that has lately taken place : the poor women who affociate
with our feamen, who were often known to perifh
in the loweft fink of human mifery, from an incurable diftemper, are faid, at prefent, to be little
afflicted

afflicted with the complaint, compared with former times.

The history of this singular disease, has exercis-
ed the pens of many ingenious physicians ; yet it
has not been satisfactorily proved, that the infection was imported by the followers of Columbus,
in his first voyage to America : on the contrary,
its trans-atlantic origin is much doubted, and the
idea that it was known before the siege of Naples,
begins now to meet with general support.

We agree with those physicians who distinguish
two diseases, which were formerly considered as
one. This opinion was first held by Dr. Duncan,
a professor of the institutes of medicine, in the
College of Edinburgh, published in the Report of
his clinical. practice, in the Public Dispensary of
that city. It has since that time received additional support from Mr. Bell's Treatise on the
Venereal Disease ; but who takes no notice of the
original author, or the arguments he had used in
favour of his doctrine. I shall therefore consider
Gonorrhœa, with a train of symptoms peculiar to
itself, as a primary disease, and incapable of producing the Confirmed Pox : the Lues, I also think,
never produces Gonorrhœa. Long and attentive
observation enables me to draw these conclusions ;
but it would be foreign from the present purpose,
to state the facts from my own experience, or that
of others.—I shall first treat of Gonorrhœa.

This

This name was given to the difeafe in the rude ages of anatomy, when the nature of the difcharge, and the parts where it arofe were not accurately known. But it is fufficiently afcertained at this day, that the running from the urethra, is only an increafe of the mucus of the parts, which becomes altered in quantity, colour, and confiftence, during the inflammation of the membrane and glands which fecrete it. We are therefore fully affured by diffections, the beft authority, that the difcharge is fuch as we have mentioned; nor does it come from excoriations or ulcers, which might alfo be fufpected, were we not convinced of the contrary.

This complaint is one of thofe, which have given full fcope to the ingenuity of impofture and empiricifm, to profit by the afflictions of mankind; hence hand-bills and news-papers teem with the accounts of wonderful cures, and *certain* prevention againft the malady. We are not furprifed to fee the credulous and ignorant, become the dupes of thefe tricks and deceptions; but it muft afford matter of aftonifhment, if not regret to fome, that there is a body of phyficians, vefted with the exclufive priviledge of preventing thefe depredations on the health of his Majefty's fubjects, who feem to have forgot that a late order in council, renewed the delegated power; to be feen in the laft editions of the Pharmacopœia Londinenfis. If any thing can prevent venereal

nereal virus from taking effect, it muſt be immediate ablution of the parts ; I apprehend nothing ſurpaſſes the finer ſoaps, and common water, which ought to be continued for ſome time, and repeated morning and evening.

٬ After impure connection, when infection has been received, it commonly appears in the following manner: a kind of tickling is felt about the end of the urethra, which at firſt may be called, rather a ,pleaſurable ſenſation, than one of pain ; but it ſoon creates uneaſineſs, particularly after the laſt drops of urine are made ; with this appears an increaſe of mucus, which is of a whiter colour, and thicker than natural ; the mouth of the urethra is of a redder complexion, is alſo wider, and ſomewhat ſwelled in the edges. A vermicular motion, at this time, is often perceived in the teſticles, and accompanied with deſire.

There is no fixing the preciſe period for the commencement of theſe ſymptoms ; nor does their early or later appearance always preſage a milder or more violent diſeaſe. They ſometimes ſhow themſelves in a few hours ; moſt generally in two or three days ; and we have repeatedly known them ſo late as ten weeks.

Theſe ſymptoms of venereal infection are occaſioned by ſome particles of the virus having come in contact with the lips of the urethra, during their ſtate of diſtenſion, and when they were

rendered

rendered exquisitely sensible by the influx of blood. When the penis is in its collapsed state, it is probable that so small a quantity would be unable to produce the effect: the poison first affects those very nicely sensitive nerves, and the inflammation of the inner membrane, and other parts, is from sympathy. There is no such thing as a fermentation taking place from the venereal matter, by which it assimilates every thing to its nature; its first action is on the nerves; and which, like a species of generation, communicates a peculiar power to the glands, by which means they generate, and pour out a matter exactly resembling the one which gave the original impulse; and like it, this matter is endowed with the property of spreading the disease. Now it requires some space of time for the poison to be in contact with the mouth of the sensible urethra, before it is sufficiently impressed to receive the disease: if, therefore, the matter which conveys the infection should be washed off, by soap and water, before the impression is finished, you will escape the complaint; and this is the whole secret of prevention.

The whole train of symptoms that follow in Gonorrhœa, are either directly from the inflammation of the parts, or its effects. The inner membrane of the urethra is more or less inflamed: certain little glands situated there, which pour out

a mucus

a mucus for defending it againſt the acrimony of the urine, partake alſo of the inflammation, and this ſo' changes their diſpoſition, that their ſecretion becomes the vehicle of the infection.

Now ſuch is the effect of inflammation here, as in all other parts, that the ſenſibility of the membrane of the urethra becomes greater than uſual.   This gives what is called the *heat of urine* ; which is felt when the urine paſſes over the tender ſurface. The pain will alſo be greater when the urine is ſmall in quantity, or when long retained in the bladder ; becauſe its more watery parts have been abſorbed, and therefore, the remainder is loaded with certain ſalts, which render it more acrid. The pain in making water is alſo greateſt, when the laſt drops come away ; this proceeds from a greater contraction in the fibres being wanted to. expel a few drops, than what is required for a . full ſtream.

The heat and redneſs of the neighbouring parts, with ſome ſwelling, are owing to the greater influx of blood, becauſe the pulſations of the arteries are increaſed ; they contract more often, and drive a greater quantity of the vital fluid through their cavities, in a given time ; hence erections in ſome ſtages of the Diſeaſe become ſo frequent and painful.—This is a ſtrong argument for avoiding all laſcivious ideas, as they have the effect of increaſing the flow of blood to parts that

H h   ought

ought to be kept collapfed and cool, which is a moft effectual means for fubduing the inflamma-tion.

When the train of fymptoms which we have defcribed above, make their appearance, they af-ford an opportunity for cutting the difeafe fhort, and preventing much pain, otherwife unavoida-ble. This may be accomplifhed by the following means :

If the patient is of a full habit, ftrong or robuft conftitution, he muft immediately fubmit to the following regimen : He is to avoid walking, rid-ing, and all kinds of exercife ; the parts muft be carefully wafhed every morning and evening in cold water, with caftile foap : abftinence from all fpiritous and fermented liquors is to be ftrictly adhered to ; and animal food of whatever kind is forbid. Weak broth, fmall-beer, water-guel, tea or coffee, and bread without butter, milk, fago, light pudding, and fuch like fare, are to form the diet. A gentle dofe of phyfic will be neceffary : boil four drams *avoirdupois* of fenna, in a half pint of water, then add an ounce of tamarinds; let them ftand a-while, then ftrain off the decoction, and drink the whole in three hours. If the patient is of a weaker habit of body, there will be little variation needed from a diet of temperance *.

* When the inflammation increafes to a great degree in a fhort fpace, bleeding is indifpenfible.

By

By way of medicine, take half a dram of arabic gum, in a half.pint of pure water, every hour or two. The intention of this is to dilute the urine: I have told you above, that the urethra, from its inflamed state, was become exquifitely fenfible: there are certain falts in the urine, which render it more acrid when little drink is taken, and in being voided over the tender and inflamed paf-fage, it gives a fenfe of heat, which is called *the fcalding of the water*. It fo happens, when a large quantity of diluting drink, fuch as gruel or barley water, is taken, that thefe falts are diffufed in a. larger proportion of fluid, and thus pafs over the inflamed membrane with lefs pain.

Along with the regimen and diluting drink, the following injections for the urethra, are to be ufed immediately on the appearance of the run-ning. The fyringe ought to contain three drams or a half ounce, made of pewter, with a conical point, fo as to fill the mouth of the urethra exactly. This being filled with one of the injections men-tioned in the margin, it muft be gently introduc-ed into the urethra, and while you hold it fteady at the point, and clofe the lips upon it with one hand, with the other you throw the injection into the paffage: the fyringe is now to be withdrawn, and the fluid to be held in for three or four minutes.

H h 2   This

This injection is to be repeated every time you make water, and as soon as that is done, before any accumulation of mucus fills the paſſage. By drinking the gum ſolution in due quantity, you will be able to inject ſo often as eight or ten times in the day, for the firſt day or two. The ſlighteſt pain ſhould be perceived in throwing up the injection; and for this reaſon, I have given three of each kind, which differ in the quantity of metallic ſalt to each. The weaker ought to be tried firſt, and the ſtronger ones in ſucceſſion : there is alſo, ſometimes, advantage in changing the one of ſugar of lead for that of white vitriol, and the contrary *.

I have

INJECTIONS.

1. Take of ſugar of lead twelve grains, pure ſpring water half a pint ; mix them.

2. Take of ſugar of lead ſixteen grains, pure water half a pint ; mix them.

3. Take of ſugar of lead twenty four grains, pure water half a pint ; mix them.

4. Take of white vitriol twelve grains, pure water half a pint ; mix them.

5. Take of white vitriol ſixteen grains, pure water half a pint ; mix them.

6. Take of white vitriol twenty grains, pure water half a pint ; mix them.

N. B. More of the ſalts, or more water may be added, if it is found neceſſary to make either of them ſtronger or weaker.

I have often known a fmart difcharge attended with confiderable pain in making water, carried off by this plan in the fpace of twelve hours: The injection muft nevertheleſs be continued for fome days after the running has difappeared, although there will be no neceffity for throwing it in fo often : the manner of living muft alfo be regular, according to the directions given above, otherwife there can be no chance of a fpeedy cure. The firft favourable fign will be the mucus becoming more thick and ropy. A fwelled tefticle, or any untoward fymptom, has never occurred in my hands from this method ; nor did I ever fee it fail in producing a cure, if duly perfifted in under the limitations prefcribed. Gentlemen who are willing to adhere rigidly to the regimen and method laid down, will foon be convinced that there is no quackery in the propofal; the procefs is fimple and intelligible, and the whole apparatus is portable. But at the fame time it ought to be remembered, that nothing is proof againft intemperance and indifcretion.

This treatment will alfo apply to the more advanced ftages of Gonorrhœa ; but it is to the recent infection alone, and the early fymptoms, that

H h 3

I would

I would wiſh you to confine yourſelves *.   I will
now direct you how to treat incipient

C H A N C R E S.

THE Chancre is a Venereal Ulcer; it ſometimes begins like a pimple with a white top,
which breaks and ſpreads. At other times it
appears at firſt merely an excoriation, as if the
ſkin

---

* A friend of mine, a wealthy merchant in London, after
ſpending the evening at a tavern in the weſt end of the town,
with ſome companions, and having drank heartily, was imprudent enough to have connection with a woman of the
town, by whom he was infected with gonorrhœa. He was a
married man, and injured his wife before he was aware of his
complaint.—In order to conceal the buſineſs, he avoided making it known to the family apothecary, but conſulted one of
thoſe ſpecious practitioners, with which the metropolis abounds.
He took much medicine, and gave large fees to his doctor,
without being at all benefited, for ſix weeks, when he became
low ſpirited and melancholy. At this time he happened to
ſee in the newſpaper my appointment to Haſlar Hoſpital. He
immediately ſet out to conſult me. His looks and appearance
alarmed me much, and I ſuſpected that his fortune, by ſome
miſchance, was ruined. He ſhortly told me his tale, as I have
related it, and I comforted him with the hopes of a ſpeedy
cure. He uſed the injections as I have deſcribed to you; the
diſeaſe diſappeared on the ſecond day; I ordered him to continue it a week or ten days longer; and he carried with him
the

skin was fretted, which soon gets deeper and wider, and never looks equal or clean on its surface. They occur most frequently on the inside of the prepuce, particularly where it joins the glans ; but occasionally also on the glans itself, and some smarting pain generally leads to the discovery. This symptom always preceds the Bubo and Confirmed Pox ; at least I have never known either without it ; although some authors tell us of the poison getting into the habit without previous ulceration. Chancres sometimes are so small and so indolent, as to give no pain ; and I have often been consulted for a Bubo, without the patient having observed any sore, till the parts were examined by myself. I have also known a small Ulcer heal up in a week, without giving any uneasiness, or having been treated as such ; and the patient was led to think that a beginning Bubo was the first symptom of the disease. On the whole, the result of my own experience is, that neither Bubo, or Lues, can be produced without preceding ul-

H h 4 ceration

---

the same for his wife ; and from that time their domestic happiness was restored. The gentleman told me since, that he had fixed his mind on either a pistol or poison, before he applied to me, as his only relief from trouble unsupportable. I do not mention this as an instance of a quick cure, because such are daily performed by others ; but it shows to what chicanery the peace of a family may be occasionally sacrificed; and it ought to be a beacon for others.

ceration in the genitals, or other parts that have been in contact with the venereal virus.

The *Chancre* is therefore a local fymptom; and if healed up in due time, will not affect the conftitution. The venereal matter is fecreted by thefe fores, it is taken up by a certain kind of veffels, which from their office are called abforbents, and by them it is carried to the glands of the groin, where it produces Buboes; or into the body, where it produces the dreadful train of fymptoms which conftitute the Lues Venerea. Now, you fee from this defcription, that the fafety of the conflitution depends upon the early treatment of a Chancre; when this is neglected, the horrors are uncalculable that muft follow: I have feen eight or ten people within the laft month, who are fuffering pains and mutilations, not to be defcribed, from having fallen into bad hands, and permitted the Chancres to fpread. We cannot fay to a certainty how long a venereal ulcer on the glans or prepuce, may remain open without endangering the abforption of the poifon; but I would not truft it beyond twenty-four hours, although we have feen them open for months, and nothing bad follow. We can, however, pronounce it as a fact, that by watching the moft early fign, after a fufpicious connection, that the danger may be prevented effectually. The parts are to be kept clean, by ablution, as directed
                                          above:

above: if any pimples are obferved, they are to be broke immediately, and wiped dry; their whole furface is then to be rubbed with vitriol of copper, or blue vitriol, as it is called, till the fpot becomes blue like the vitriol: if there are any excoriations, let them alfo be rubbed over, and apply a little furgeon's lint, very thinly, over the whole. The lint may be renewed after the ablution, morning and evening, and the blue fpots will fall off fpontaneoufly in floughs, and leave a clean ruddy furface beneath, which needs no application but the lint, as thin as it can be laid over, and the fame attention to cleanlinefs. If, however, the furface of the fore fhould lofe its ruddy appearance, and not clofe quickly after the feparation of the flough, the vitriol may be renewed; but this will feldom be needed. There is fometimes a neceffity of touching eight or ten little fores in this manner; a trifling inflammation follows, but never fuch as to give alarm. It will therefore be proper to follow the regimen directed in Gonorrhœa, as it will accelerate the cure.

It has long been the cuftom of deftroying Chancres by cauftic, as you will fee in medical books; but the word itfelf, to perfons not converfant with the fubject, implies a painful operation. I have tried every kind, and I am now fatisfied that the blue vitriol is the beft. It is

the

the nature of a venereal fore to be obſtinate; the ſurface never looks clean, and it never ſhows granulations, or little ſprouts of new fleſh, which form the character of ulcers diſpoſed to cloſe and heal: we therefore employ a cauſtic, which de-ſtroys their ſurface altogether to a certain depth; this is gradually ſeparated from the ſound fleſh below, and a new ſore is formed, that has nothing venereal in its nature, and which heals in thirty-ſix hours, if not large. If, therefore, you care-fully watch the firſt appearance of infection, as I have directed, you will certainly ſecure your con-ſtitution; the ſurface of the Chancre being ſmall, it is likewiſe ſuperficial.

If the Chancre is of ſome days ſtanding, we can-not anſwer for the ſafety of the conſtitution; but even in that caſe, I would recommend the ſame treatment. It is always neceſſary to check their ſpreading, which is ſometimes inconceivably rapid. Some authors, of great note, ſay, that in this ſitua-tion, the cauſtic haſtens the abſorption; but our experience does by no means juſtify this con-cluſion: even if it were ſo, it could be no argu-ment againſt the ſpeedy cure; for while it re-mains open, it muſt generate venereal matter, which is every moment in danger of being carried into the blood. I would therefore adviſe the ap-plication of the blue vitriol to every Chancre, and

consult

confult with fome phyfician or furgeon, on the propriety of ufing mercury for the fecurity of the conftitution.

A folution of blue vitriol, is often ufeful in the cafe of fores, between the prepuce and glans, when the fkin, from inflammation or ftricture, cannot be drawn back. Two grains will be ftrong enough for an ounce of water; it may be thrown between the fkin and glans with a fyringe, morning and evening, and care muft be taken it does not get into the urethra. Thefe fores are known to be prefent, by the pain occafioned by fqueezing the glans, or moving the prepuce, and alfo, by the matter flowing from between thefe parts.

C A S E

OF

AMPUTATION AT THE SHOULDER,

In a wounded French Officer,

Performed at Sea by Mr. BURD, *Surgeon of His Majesty's*
Ship NIGER.

April 26th. L AST night, Jean Moerieton, a French officer, apparently a healthy vigorous man, about twenty-four years of age, was brought on board here, having received a wound with a mufquet ball in the right fhoulder. Upon examination, I found the ball had entered at the deltoid mufcle, about two inches above its infertion, and paffed out at the fuperior part of the fcapula. I dilated the anterior wound, and readily difcovered the neck of the os humeri to be fractured, and much fplintered. I alfo was enabled to feel, with the firft

finger

finger of my right hand, introduced at the wound, that its head was fhattered into two or more pieces; a fracture of the fcapula, where the ball paffed out, was very, evident; the clavicle was thrown fo much upwards, as totally to prevent my being able to diftinguifh the firft rib above it. Having afcertained the nature of the wound, I confidered the operation as indifpenfable, but at the fame time refolved to defer it till the firft inflammation fhould fubfide, and then perform it; therefore nothing more was done the firft night, than to cleanfe the wounds, apply light dreffings, and a proper bandage. An anodyne was given him, which procured a tolerable night. On infpection this morning, the anterior wound was found to have bled confiderably; the dreffings were removed, but no artery could be difcovered, to be fecured. The wound was now dreffed again as laft night, and as there was a great deal of tenfion about the parts, a cataplafm was applied.— Repet. hauft. anodyn. h. f.

May 1ft. Has had tolerable reft thefe four nights paft, with the ufe of anodynes; p. little quickened; b. bound; eats fago for dinner, with a little wine in it.—Bib. aq. hord. Wounds are dreffed daily; have a very irregular lacerated appearance, and difcharge copioufly a highly

fœtid

fœtid pus, of pretty good confiftence.—Cont. cataplafm.

May 3d. As the inflammation was nearly, if not entirely gone, and a very copious difcharge of pus continued, by which the patient was much debilitated, at the fame time the weather fo moderate, that the fhip could be kept pretty fteady, I determined to perform the operation to-day. When he was placed upon the table, which was in the gun-room, I endeavoured to make compreffion on the artery as it paffes over the firft rib, but without fuccefs; for not only the clavicle's being thrown upwards, but alfo fome degree of tenfion, prevented its being effected, except very partially. I now attempted to make the compreffion in the axilla, but the head of the humerus being fo much fhattered, yielded to the leaft preffure; finding both fail, I felt myfelf very unpleafantly fituated; but without the operation's being performed, it was evident that death was the certain confequence to the patient; therefore refolved to give him the chance.

Having every thing prepared, I appointed a French furgeon (who was furgeon of the veffel the patient was wounded in) to make what compreffion he could upon the artery above the clavicle, which, as mentioned before, was only partial. Mr. Brown, furgeon's

furgeon's mate of the Niger, ſtood at hand, to
give me the neceſſary inſtruments, apply ligatures,
&c. The arm was now ſtretched out, and ſup-
ported at nearly a right angle with the body, the
ſhoulder projecting over the ſide of the table.
A circular inciſion was now made through the
ſkin and cellular ſubſtance, about the inſertion of
the deltoid muſcle, into the humerus, before pro-
ceeding further, any divided blood-veſſels were
ſecured. The teguments retracted about three
quarters of an inch. At the edge of the retracted
teguments, on the inner and under parts of the
arm, I applied the knife, and divided the muſcles
down to the bone, all round, except a portion in
which the humeral artery was included: any large
blood-veſſels that were divided by this inciſion
were ſecured. In the undivided portion of muſcle,
I diſtinctly felt the pulſation of the artery; upon
which I placed the thumb of my left hand; and
then finiſhed the diviſion of the muſcles down to
the bone: upon the artery being divided, a pro-
digious flow of blood followed. (Had the French
ſurgeon been of much ſervice, he now forſook
me, and removed the little compreſſion he was
making.) I was very ſoon enabled to paſs a ligature round it, and ſecured it, fortunately without
the loſs of ſo much blood as might have been
expected.

expected, when we confider the fize of the veffel, its vicinity to the heart, and almoft a total want of compreffion, for my thumb effected only a partial one ; as it was fecured with the needle, the nerve was inevitably included in the ligature. (During the whole of the operation, Mr. Brown gave me very great affiftance, but particularly at this part of it.)   I now fecured every blood veffel; for even the fmalleft bled freely, after the larger were fecured.   With a ftrong round-edged fcalpel, I made a perpendicular incifion down to the bone, beginning at the lower part of the wound (which I had dilated upwards the firft night he came on board) and terminating in the circular incifion, about an inch and a half on the outfide of the humeral artery, the bleeding veffels were imme-diately fecured.   Finding the deltoid mufcle was in a gangrenous ftate, I made the upper flap about one third fmaller than the lower, and proceeded to feparate the flaps from the os humeri ; which being effected, the arm came away, not leaving half an inch of the humerus attached to its head. I now faw that the head of the humerus was fhivered into various pieces of different fizes. After cutting the capfular ligament all round, it was, with fome difficulty, I removed the firft piece ; the others came away eafily, accompanied with

- 5

with a part of the glenoid cavity.   I found two
pieces of the fcapula detached, which were dif-
fected out ; one of them was the greateft part of
the fpine: the removal of all thefe, together with
the gangrened flefh, protracted the operation to a
great length of time, which the patient bore with
aftonifhing fortitude.   In the courfe of the operation, eight or ten veffels were fecured, fome with
the needle, others with the tenaculum.   After
cleanfing the furface of the flaps, and applying a
fecond ligature upon the humeral artery, about
one quarter of an inch above the firft, they were
brought in contact, and adhefive ftraps applied to
retain them together, leaving an aperture in the
moft depending part, fufficient to difcharge the
remaining very fmall pieces of bone, which could
not be removed by the knife.   After the ftraps,
pledgets of cerate, or lint, were applied, and a
comprefs of lint and tow, with a flannel roller over
all.—An anodyne of tinct. opii. gt. 40. was given
immediately, and repeated at bed-time.

May 5th.   The pofterior wound was dreffed to-
day ; difcharge copious ; the patient complains of
almoft conftant naufea ; he has alfo fome fpafmo-
dic affections, frequently ; no ftool fince the operation ; in the afternoon he complained of pain
in the lower belly, and inability to pafs urine,

with frequent tendency: the pulfe quickened; fkin hot; great thirft; with fome head-ach.— Warm fomentations were applied to the belly, and he foon paffed his urine freely.—Hab. Enem. Commun.—A faline draught was given him every fecond or third hour, and an anodyne at night.— Dreffings have not yet been removed from the ftump.

May 7th. Soon after the operation of the in- jection, moft of the unfavourable fymptoms were confiderably relieved; and, by continuing the fa- line mixture, went intirely off. This morning there was a recurrence of the pyrexeal fymptoms. I removed the dreffings from the ftump, and put frefh applications in their place; foon after which the febrile fymptoms fubfided. The flaps were found to have adhered no where; their furface have a very floughy appearance. Adhefive ftraps we applied to the fuperior part of the flaps, fo as to keep them flightly in contact: the inferior part was lightly dreffed. As fuppuration had already begun to take place, over all a comprefs of flannel roller was applied. The pofterior wound looks well, and difcharges a laudable pus; it is dreffed daily. Sago, with a little wine, is given him for dinner: drinks water, with a fmall quantity of Oporto wine in it.—Omitt. Hauft. Anodyn.

May 9th.

May 9th. About an inch and a half of the lower part of the upper flap is in a gangrenous ftate; difcharge very copious; extremely fœtid, but of pretty good confiftence; p. calm and weak; fkin moift; b. rather coftive; refts but indifferently without an anodyne. Repet. Hauft. Anodyn. h. f. ut antea. Eats a little mutton for dinner; and in the courfe of twenty-four hours drinks nearly a bottle of Oporto wine, in water: he is dreffed now twice a day.

May 11th. Difcharge continues very great and fœtid; p. regular; feveral of the ligatures have fallen off; his diet and drink, as mentioned laft report, continued. Omitt. Hauft. Anodyn.

May 14th. Part of the fuperior flap, that was in a gangrened ftate, fell off to-day: the lower flap looks extremely well, and granulates kindly; difcharge continues copious; all the ligatures, except that of the humeral artery, have been re-moved. The patient has great fpirits; general health good; and fleeps very well, without the ufe of an anodyne.

May 16th. Granulations are fhooting out from every part of both flaps; difcharges a well di-gefted pus, not fo profufe, and little fœtor. This morning I accompanied him to Forton Hofpital, where

where he was left under the care of Mr. David
Patterfon, furgeon there.

Aug. 17th. He was difcharged perfectly well,
and returned to France.

WILLIAM BURD.

Niger, October 14, 1796.

POSTSCRIPT.

Now that my labours are, for the prefent, brought to a conclufion, I cannot let them efcape from my hands without emotions of a particular kind. I ftand in a predicament, different from all authors which have preceded me in the fame line; for I am doomed to remain on the Spot of Action. If a fallacious ftatement of occurrences has been purfued, it will be quickly expofed. As I have on all fubjects thought for myfelf; fo the freedom of opinion will be retorted: and as fome innovations have taken place, I muft expect that they will be examined with fufpicion, and criticifed with diftruft. While the approbation of the world is an ingredient in the happinefs of an individual, it is in vain to oppofe to thefe, purity of intention, or rectitude of conduct. Hence my reflections muft be checquered with hopes and fears, or clouded by mifconftruction and dif-appointment.

It will continue to be my tafk, to improve this Work, as future experience may fupply materials,

I   and

and to correct what may be found exceptionable; I shall therefore receive with gratitude every communication, which the Surgeons of the Navy may think me worthy.

———————

To the younger Members of the Profession in the Navy, I have a short advice to offer. As study and diligence lead to preferment in every department of life, so in our line, they have their rewards. We fill a most important station in the service of our Country; nay of much more importance than we can expect credit for, because many of our best actions must sleep with ourselves, as medical abilities are not to be appreciated by common observers. Yet this very circumstance is a stimulus to exertion; for it keeps alive the spirit of perseverance, from the hope that merit will at last be discerned, and meet with success. Changes have lately happened in the medical department, that have multiplied the posts of honour; and from the abilities that now manage the administration of science, over us, the ingenious and active student cannot fail to be distinguished and known. But as the Board of Physicians will applaud and protect talents and worth, they have it in their power to detect and chastise ignorance and sloth. Of this you must seriously reflect; for to incur displeasure, is little

short

fhort of lofs of reputation. Our defections of duty carry with them a higher degree of reprehenfion, than are prefixed to fome other employments; becaufe not only the lives of individuals are intrufted to our care, but the health of a Fleet, on which the fafety of Great Britain may depend, may devolve upon us. This was really the cafe, during the extenfive Contagion, fpread from the French prifoners, and during the general Scurvy, in the fpring and fummer of 1795. As our deferts, therefore, are not always known, fo our faults cannot always be concealed.

I cannot help congratulating the fervice on the arrangements and appointments that have taken place at the hofpitals. They promife to hand down to pofterity, that fpirit of improvement which is effential to public welfare : Thus end the Naval Occurrences of 1796.

F I N I S.

www.ingramcontent.com/pod-product-compliance
Lightning Source LLC
Chambersburg PA
CBHW020857210326

41598CB00018B/1699